Chemical Bonding

Elementary Chemistry for Teens

Copyright © 2021 Kok Koon Leong

All rights reserved.

No part of this book may be reproduced in any form or by any electronic or mechanical means, including information storage and retrieval systems, without written permission from the author.

Contents

Chapter	Topic	page
1	Foreword	i
2	"You Complete Me"	1
3	The Periodic Table and Valency	8
4	Test Your Understanding 1	12
5	The Ionic Bond	15
6	Test Your Understanding 2	22
7	The Covalent Bond	25
8	Giant Covalent Structures	33
9	Comparing Ionic and Covalent Substances	37
10	Writing Formulas of Ionic and Covalent Compounds	38
11	Test Your Understanding 3	44
12	Metallic Bonding	48
13	Hybrid Bonding	50
14	Practice Questions	51
15	Answers to Questions	55
16	The Periodic Table of Elements	62

Foreword

This short guide is written in line with the Singapore GCE O Level syllabus (for 15 to 17-year-old students) Pure Chemistry and Combined Science Chemistry.

The aim of this guide is to present to you dear student, concepts, and principles to help you understand the topic of Chemical Bonding.

To get the maximum benefit from this guide,

1. Read each chapter carefully.

2. Practice the questions in the 'Test Your Understanding' chapters. These are short questions which help you to apply what you have learned. You can print the exercises as hardcopies and write on them if you have ordered the hardcopy book. Or you can read the questions on your computer and write your answers on a piece of paper. Then, only after you have attempted the questions, check your answers in the Answers chapter.

3. When you read a chapter, keep a pencil or pen and a piece of paper (or jotter book) near you. Sit comfortably at your study table and when you come across an important fact or principle or a question that you want to ask, jot it down. Writing things down helps you to remember and understand what you read.

4. Once you have sufficient understanding of the chapters, be ready to teach someone else. Teaching others helps you become better. You will be amazed at how much you can improve when you teach others.

All the best on your road to success!

Kok Koon Leong

"You Complete Me"

Let's examine the elements in the noble gases. Below is the group of noble gases as seen in the Periodic Table.

2
He
Helium
4
10
Ne
Neon
20
18
Ar
Argon
40
36
Kr
Krypton
84
54
Xe
Xenon
131
86
Rn
Radon
-

Helium, the first element of the noble gases, has 2 protons (2p), 2 neutrons (2n), and 2 electrons:

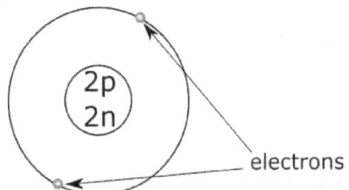

The helium atom has 2 electrons in its only shell. This is the maximum number of electrons that this shell can hold. We see that the shell is 'full' (of electrons).

Neon and Argon the next elements in the group, have the following structures:

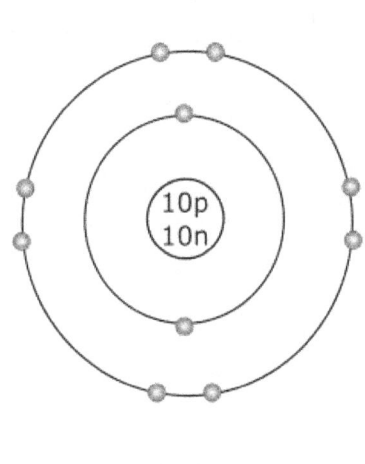
neon

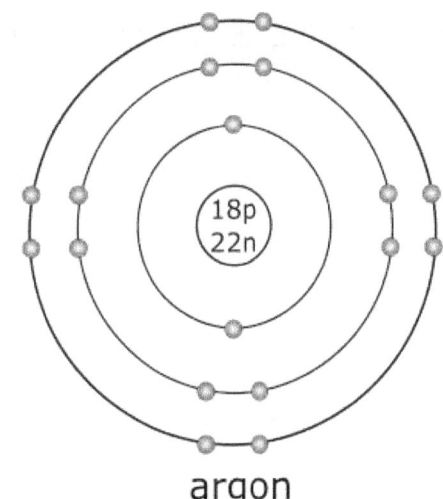
argon

Notice that both neon and argon have 8 electrons in the outermost electron shells (also known as **valence shells**).

And, krypton, xenon and radon, the remaining elements in the group, all have 8 electrons in their valence shells (the electrons in the valence shells are called 'valence' electrons – no surprise there!).

For atoms, 8 is a lucky number. Once a valence shell has 8 electrons (or in the case of helium, 2 electrons), an atom is 'satisfied'. It does not 'want' to add any more electrons to this valence shell. So, when that happens, we say that the valence shell is **complete** (i.e., it has 8 electrons).

Now, think of any compounds that have helium, neon, argon or any of the elements in the noble gas group as one of their constituents …

Take a minute. Or two. Or three.

Okay, there are a few (if you google it), but these are not the usual compounds that one comes across. In general, **noble gases do not form compounds readily**. That's why they are called 'noble' gases. In ancient times, members of the nobility do not mix with commoners. In general, noble gases are so called because they do not react readily with the other elements.

Let's summarize our observations:

1. Helium has a **complete valence shell (of 2 electrons)**.
2. Neon, argon and all the other noble gases have **complete valence shells (of 8 electrons)**.
3. To achieve stability, the valence shell should ideally have 8 electrons (except for helium which achieves stability by having 2 electrons)

Nobel Gases as 'Role Models'

Why are we looking at the noble gases? Isn't this book about Chemical Bonding?

Well, the noble gases are role models for all the atoms in the Periodic Table.

Every atom in the Periodic Table tries to emulate the noble gases electronic configuration, especially that of their valence shells. Look at the example below:

| 3 Li Lithium 7 | 4 Be Beryllium 9 | 5 B Boron 11 | 6 C Carbon 12 | 7 N Nitrogen 14 | 8 O Oxygen 16 | 9 F Fluorine 19 | 10 Ne Neon 20 |

(with 2 He Helium 4 above Neon)

In the picture above, you can see that the boxes containing helium (He), lithium (Li), beryllium (Be) and boron (B) are coloured red while those of nitrogen (N), oxygen (O), fluorine (F) and neon (Ne) are green. For the time being let's leave out carbon (C), the odd man out. Let the elements in the red boxes be the Red Team and the elements in the green boxes be the Green Team.

Look at the electronic structure or the Red Team:

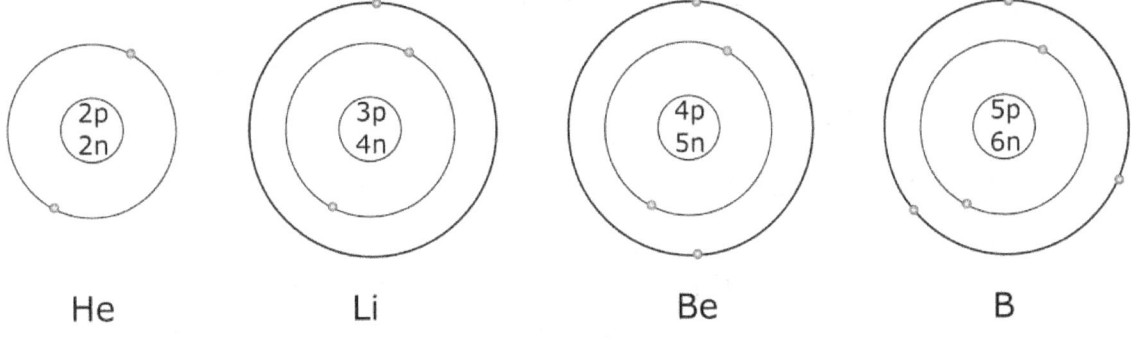

He　　　Li　　　Be　　　B

As helium is the role model, every member of the Red Team, i.e., Li, Be, and B tries to be like He to achieve stability.

How do they do this?

Simple. They lose their valence electrons!

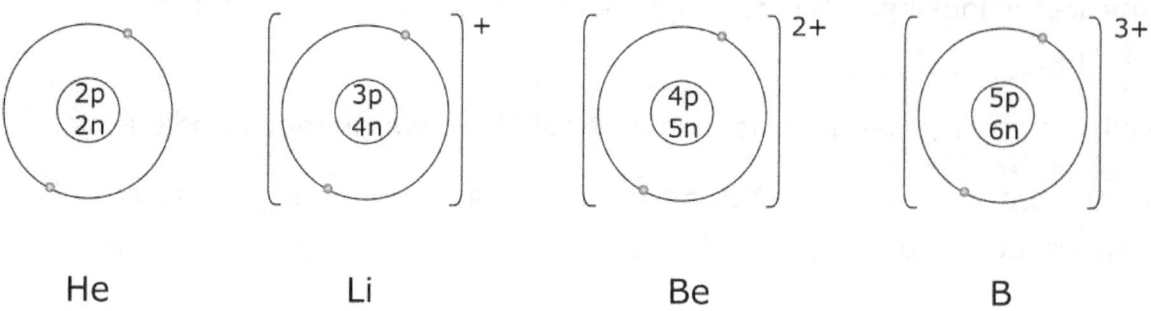

See the picture above? Each atom of the Red Team now has exactly the same electronic configuration as helium, i.e., each of them has exactly 2 electrons now. Just like helium.

But, what about the square brackets enclosing the lithium, beryllium and boron atoms? And the '+', '2+' and '3+' at the top right-hand corner of the brackets?

Look at the lithium (Li) atom. It has 3 protons and 2 electrons. Protons and electrons are electrically charged particles. **A proton has a single positive charge while an electron has a single negative charge.** Initially, the atoms, Li, Be, and B, all have equal number of protons and electrons, so the net charge for each atom is zero, but now, having lost all their valence electrons, lithium, beryllium and boron do not now have zero charge:

Li has 3 protons and 2 electrons, so its net charge is (+3) + (-2) = +1.

Beryllium has 4 protons and 2 electrons, so its net charge is (+4) + (-2) = +2.

Boron has 5 protons and 2 electrons, so its net charge is (+5) + (-2) = +3.

Look at the numbers. Aren't they the same as those on the top right-hand corner of the square brackets?

This is the net charge of each of the atom in the picture.

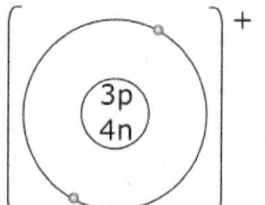

So, lithium now has a net positive charge and we can either draw the lithium atom *sans* its valence electron as shown on the left, or we can write it as Li^+.

Atoms, where the number of protons is not equal to the number of electrons, are called **ions**.

Thus, the beryllium ion can be written as Be^{2+} and the boron ion as B^{3+}.

Let us now examine the Green Team:

7	8	9	10
N	O	F	Ne
Nitrogen	Oxygen	Fluorine	Neon
14	16	19	20

The Green Team's atomic structures would look like this:

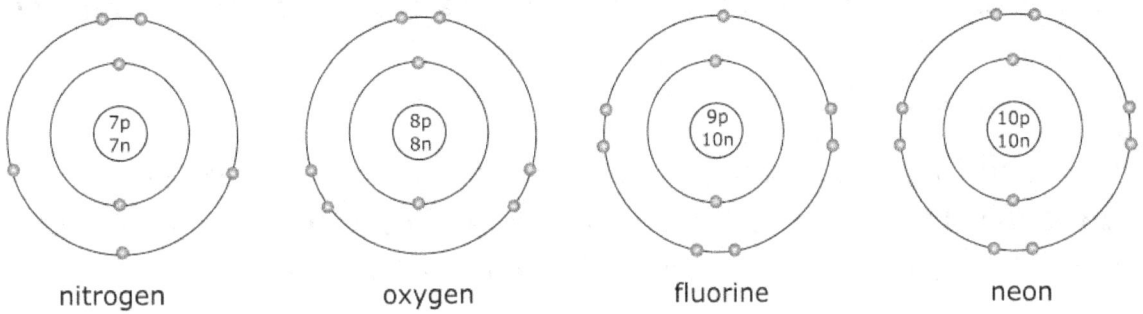

Since neon is the 'role model' of this team, the other elements try to emulate neon's electronic structure. They do so by gaining extra electrons in their valence shells.

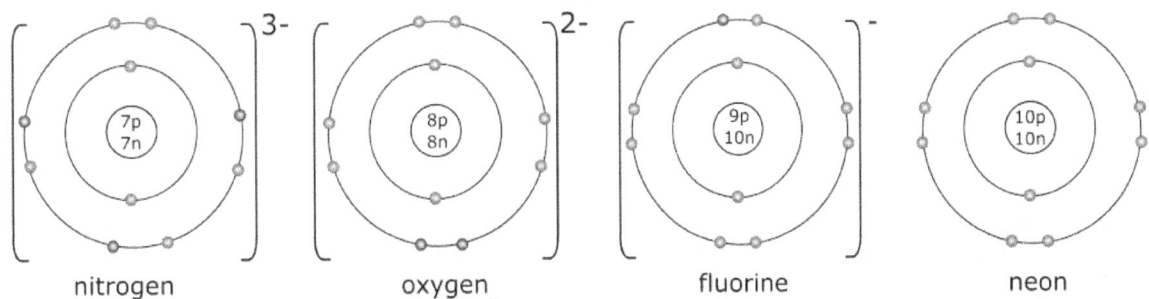

See the blue balls in the picture above? They are electrons that each atom (nitrogen, oxygen and fluorine) has gained.

Now all the atoms have the same electronic configuration as neon.

Nitrogen now has 3 more electrons than protons (+7) + (-10) = -3

Oxygen now has 2 more electrons than protons (+8) + (-10) = -2

Fluorine now has 1 more electron than protons (+9) + (-10) = -1

To sum up, consider the picture below:

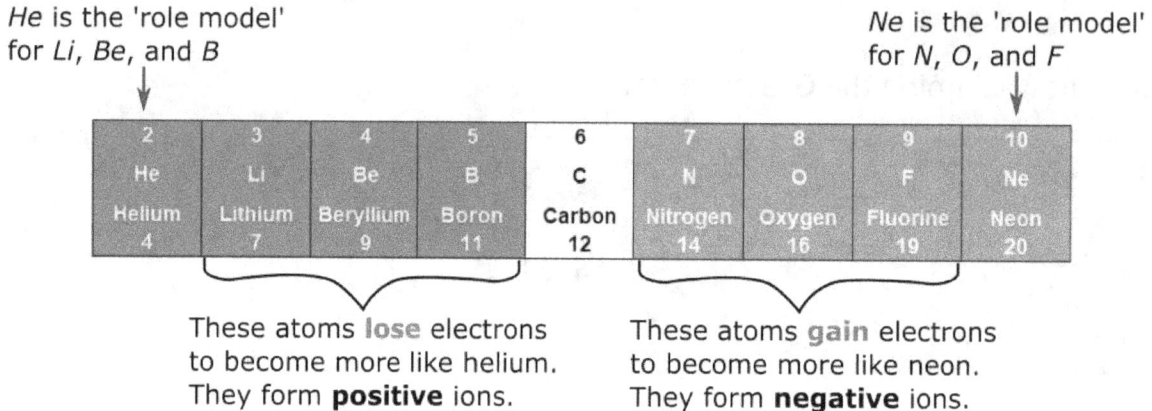

These atoms **lose** electrons to become more like helium. They form **positive** ions.

These atoms **gain** electrons to become more like neon. They form **negative** ions.

Why do atoms behave like this? **Atoms are always trying to achieve stability by losing or gaining electrons.** The electronic configurations of noble gases make them very stable.

Lithium, beryllium and boron are metals and all **metals try to achieve stability by obtaining the electronic configuration of the noble gas before them** in the Periodic Table, i.e., from the period just above them.

Nitrogen, oxygen and fluorine are non-metals and all **non-metals achieve stability by obtaining the electronic configuration of the noble gas on the same period as them.**

The picture below shows part of the Periodic Table:

3	4
Li	Be
Lithium	Beryllium
7	9
11	12
Na	Mg
Sodium	Magnesium
23	24

					2
					He
					Helium
					4
5	6	7	8	9	10
B	C	N	O	F	Ne
Boron	Carbon	Nitrogen	Oxygen	Fluorine	Neon
11	12	14	16	19	20
13	14	15	16	17	18
Al	Si	P	S	Cl	Ar
Aluminium	Silicon	Phosphorus	Sulfur	Chlorine	Argon
27	28	31	32	35.5	40

The metals (in bold) Li, Be, B lose electrons to achieve the electronic configuration of He while Na, Mg and Al – all metals – lose electrons to achieve the electronic configuration of Ne.

The non-metals, N, O and F gain electrons to achieve the electronic configuration of Ne while P, S and Cl (all non-metals) gain electrons to achieve the electronic configuration of Ar.

The charge of an ion is also known as its **valency**.

An easy way to figure out the valency of an ion is as follows:

Consider two atoms X and Y:

If X loses 2 electrons, then you get X^{2+} ion

If Y gains 2 electrons, then you get Y^{2-} ion.

If X loses 3 electrons, you get X^{3+} ion.

If Y loses 3 electrons, you get Y^{3-} ion.

In short, **every time an atom loses electrons, it becomes a positive ion.**

If the atom gains electrons, it becomes a negative ion.

Representing ions in writing.

You will notice that the sodium ion has a single net positive charge and is represented as Na^+, and not Na^{1+}. The fluoride ion is also represented as F^- and not F^{1-}.

This is the convention in representing the ions in writing. The number '1' is not written. All other numbers are written, for example, Al^{3+} and N^{3-}.

The Periodic Table and Valency

A portion of the Periodic Table is reproduced below.

I	II		III	IV	V	VI	VII	0
								2 He Helium 4
3 Li Lithium 7	4 Be Beryllium 9		5 B Boron 11	6 C Carbon 12	7 N Nitrogen 14	8 O Oxygen 16	9 F Fluorine 19	10 Ne Neon 20
11 Na Sodium 23	12 Mg Magnesium 24		13 Al Aluminium 27	14 Si Silicon 28	15 P Phosphorus 31	16 S Sulfur 32	17 Cl Chlorine 35.5	18 Ar Argon 40

Look at the vertical columns in the table. Each column has been given the same colour.

The Roman numerals at the top of each group shows its group number.

Let's look at the two Group I elements as shown in the Periodic Table above:

Element	Proton/ Atomic Number	Electronic Configuration	Symbol of ion	Charge of ion
Li	3	2,1	Li^+	+1
Na	11	2,8,1	Na^+	+1

Similarly, we can construct a table for the first two Group II elements:

Element	Proton/ Atomic Number	Electronic Configuration	Symbol of ion	Charge of ion
Be	4	2,2	Be^{2+}	+2
Mg	12	2,8,2	Mg^{2+}	+2

Similarly, Group III elements would have a charge of +3 for each ion.

Element	Proton/ Atomic Number	Electronic Configuration	Symbol of ion	Charge of ion
B	5	2,3	B^{3+}	+3
Al	13	2,8,3	Al^{3+}	+3

What about Group V elements?

Element	Proton/ Atomic Number	Electronic Configuration	Symbol of ion	Charge of ion
N	7	2,5	N^{3-}	-3
P	15	2,8,5	P^{3-}	-3

And Group VI elements ...

Element	Proton/ Atomic Number	Electronic Configuration	Symbol of ion	Charge of ion
O	8	2,6	O^{2-}	-2
S	16	2,8,6	S^{2-}	-2

The table below shows the first two Group VII elements:

Element	Proton/ Atomic Number	Electronic Configuration	Symbol of ion	Charge of ion
F	9	2,7	F^-	-1
Cl	17	2,8,7	Cl^-	-1

We seemed to have missed the Group IV elements, didn't we? Well, Group IV elements do not form ions easily, so we will leave them out for the time being.

Let's summarize the charges of the ions going across the Periodic Table from left to right.

Group	I	II	III	V	VI	VII
Charge of ion	+1	+2	+3	-3	-2	-1

We notice a few things:

1. **Groups I to III are metals**, while Groups V to 0 are non-metals.
2. **Metals form positive ions, while non-metals form negative ions.**
3. The charge of a metal ion is the same as its group number.

4. The charge of a non-metal ion is equal to its group number minus 8. For example, the charge of a Group VI element is 6 – 8 = -2.

Test Your Understanding 1

Write your answers in the spaces below the question.

1. (a) Complete the outline diagram shown below, to show the electronic arrangement of a chlorine atom. The electrons of the first shell, represented below as crosses, has been drawn for you.

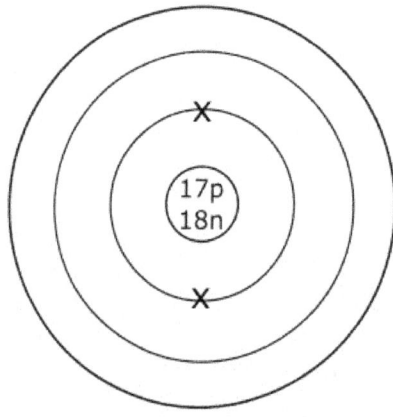

(b) In the space below, write down the symbol for an ion of chlorine.

(c) How many electrons does an ion of chlorine have?

(d) What is the name of the noble gas that has the same electronic configuration as this ion of chlorine?

2. The symbol for magnesium is $^{24}_{12}Mg$.

(a) Write down the number of subatomic particles in an atom of magnesium in the table below.

Sub-atomic particles	Number
Proton	
Neutron	
Electron	

(b) How many valence electrons does this atom of magnesium have?

(c) Write down the symbol for an **ion** of magnesium in the space below.

(d) Why does the atom of magnesium form this type of ion?

3. Use the Periodic Table to complete the following table.

Element	Electronic configuration / structure of atom	Symbol of ion formed	Name of ion	Electronic configuration / structure of ion	Noble gas whose configuration of ion is similar to
Calcium			calcium ion		
Fluorine	2,7	F$^-$	fluoride ion		
Lithium					
Chlorine			chloride ion		
Nitrogen					
Potassium					

The Ionic Bond

We have just seen that atoms gain or lose electrons to achieve the electronic configuration of noble gases, i.e., they become stable.

So, imagine an atom of chlorine, with an electronic structure of 2, 8, 7.

It would need to gain one more electron to give it the configuration 2, 8, 8 to be stable.

Where would this single electron – that the chlorine atom needs – come from?

It gets it from another atom, in this case, an atom that needs to lose a single electron. For example, a sodium atom.

So, if these two different atoms are in each other's vicinity, the sodium atom would transfer an electron to the chlorine atom to form a sodium ion and a chlorine ion:

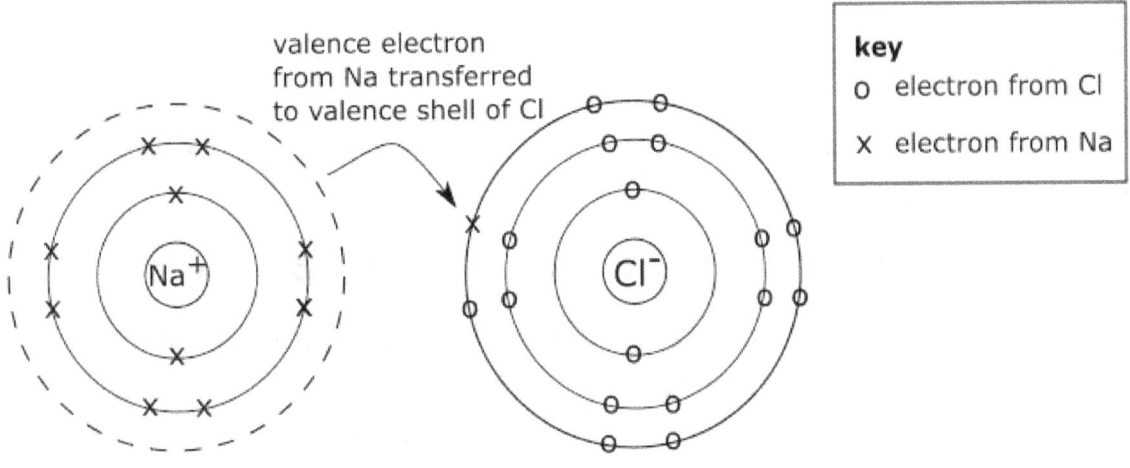

The sodium atom loses an electron (e^-) to become a positive ion:

   Na → Na^+ + e^-

The chlorine atom gains this electron to become a negative ion:

   Cl + e^- → Cl^-

Notice that we used e⁻ to represent the electron.

We now have two ions, one positive and one negative ion next to each other. Positive and negative ions attract and a very strong **electrostatic force of attraction** is thus established between the two oppositely charged ions. A new substance is formed when sodium reacts this way with chlorine.

This new substance is called sodium chloride. We all know it as common salt, used in cooking to give taste.

We can represent the sodium chloride compound like this:

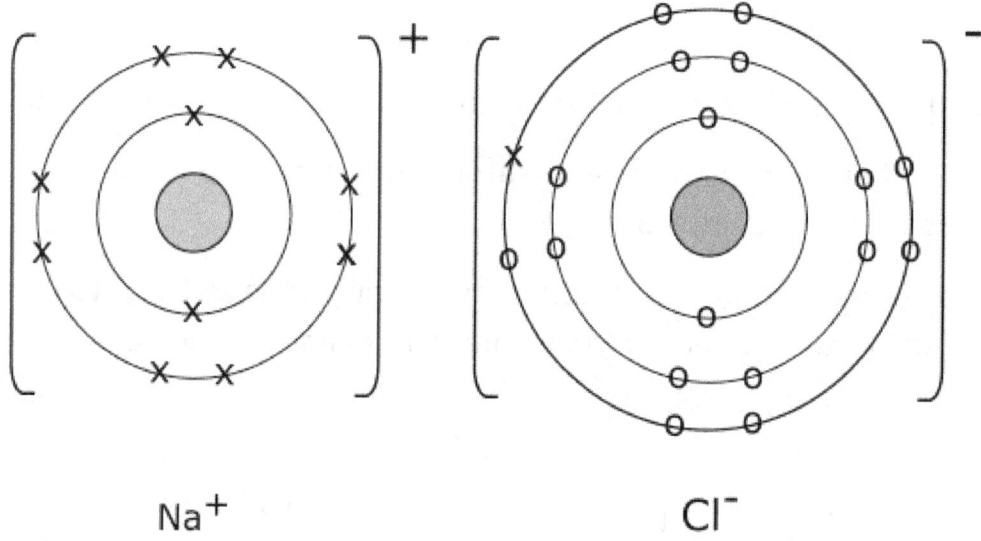

Na⁺   Cl⁻

This kind of bonding, where the electrons from one ion is transferred to another, is called **ionic bonding**.

The substance formed from ionic bonding is called an **ionic compound**.

Note that we have used 'x' to represent the electron from the sodium atom and 'o' to represent the electron from the chlorine atom. This is to show how the electron is transferred from sodium to chlorine. In reality, electrons are identical no matter which atom they come from.

Let us look at another example.

Consider the magnesium and oxygen atoms with electronic configurations of 2, 8,2 and 2,6 respectively.

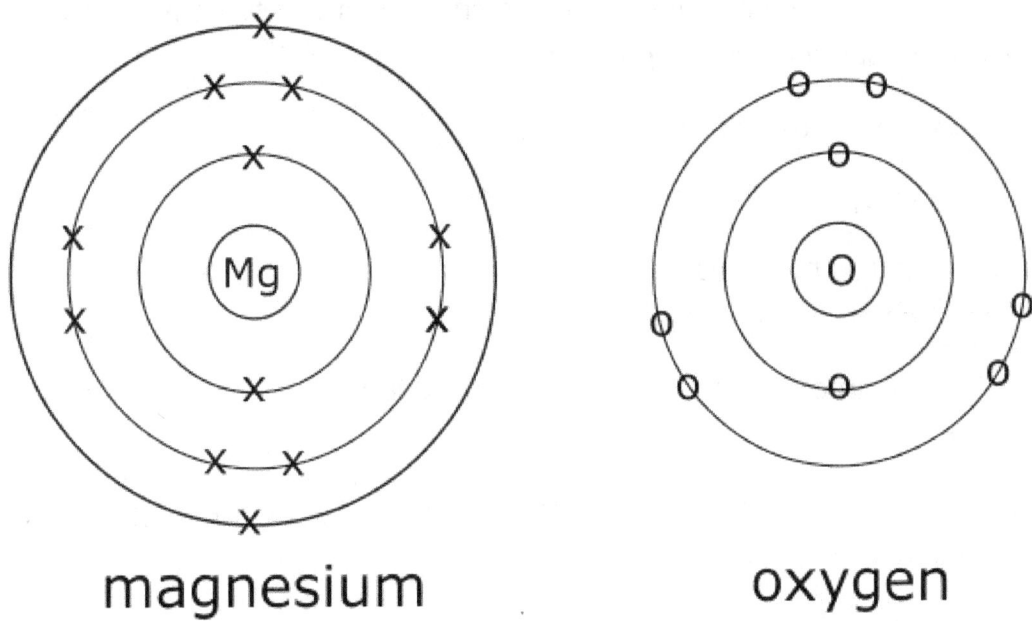

magnesium   oxygen

To gain a stable electronic structure, it transfers its two valence electrons to oxygen which also gains a stable electronic structure when it accepts these two electrons. Both gain the electronic structure of neon.

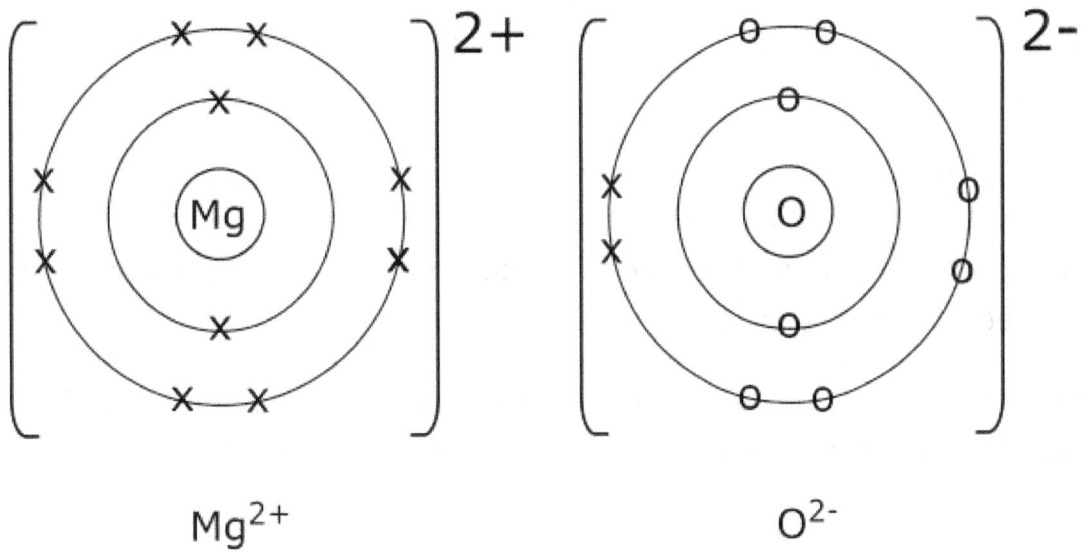

Mg^{2+} O^{2-}

Because magnesium loses 2 electrons, it becomes a double positively charged ion, Mg^{2+}. The oxygen atom, having gained 2 additional electrons, becomes a double negatively charged ion, O^{2-}. And, because of their opposite charges, these two ions attract each other strongly. We say that there is a strong **electrostatic force of attraction between the oppositely charged ions**.

We have so far looked at ions having opposite charges of the same magnitude. What if the ions involved have charges of different magnitudes?

Consider the compound magnesium chloride.

It is formed when magnesium combines with chloride ions.

As we can see from the previous two examples, the magnesium ion has a +2 charge while the chloride ion has a -1 charge.

Now **a compound formed from oppositely charged ions must always have a zero net charge.**

So, it would take two chloride ions to neutralize the charge of one magnesium ion, i.e., 2(-1) + (+2) = 0.

Hence, we can draw the ions of magnesium chloride like this:

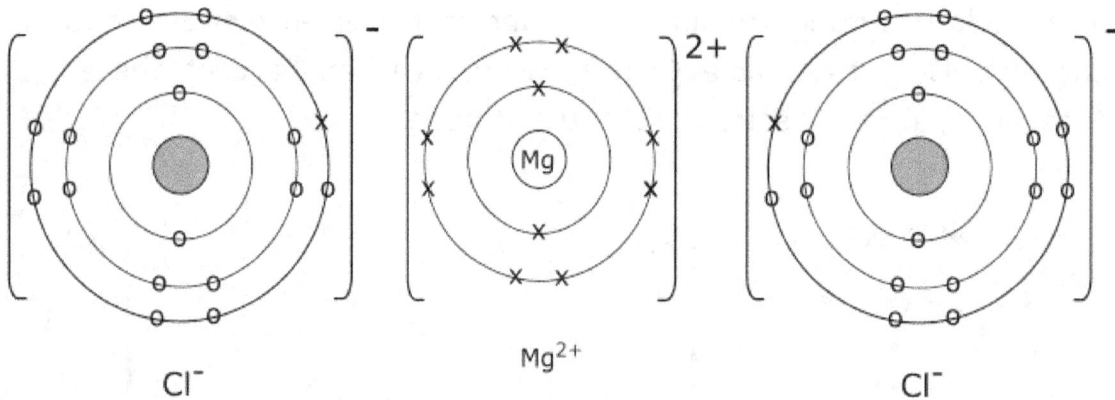

Cl⁻ Mg²⁺ Cl⁻

In all the examples that we have seen, did you notice that ionic compounds are formed between metals and non-metals?

Yes, generally, **ionic bonds are formed between metals and non-metals**. There are some exceptions but these are beyond the scope of this guide.

Lattice Structure

You might get the idea from the preceding pages that ionic compounds such as sodium chloride exist like this, i.e., as individual NaCl pairs close to each other :

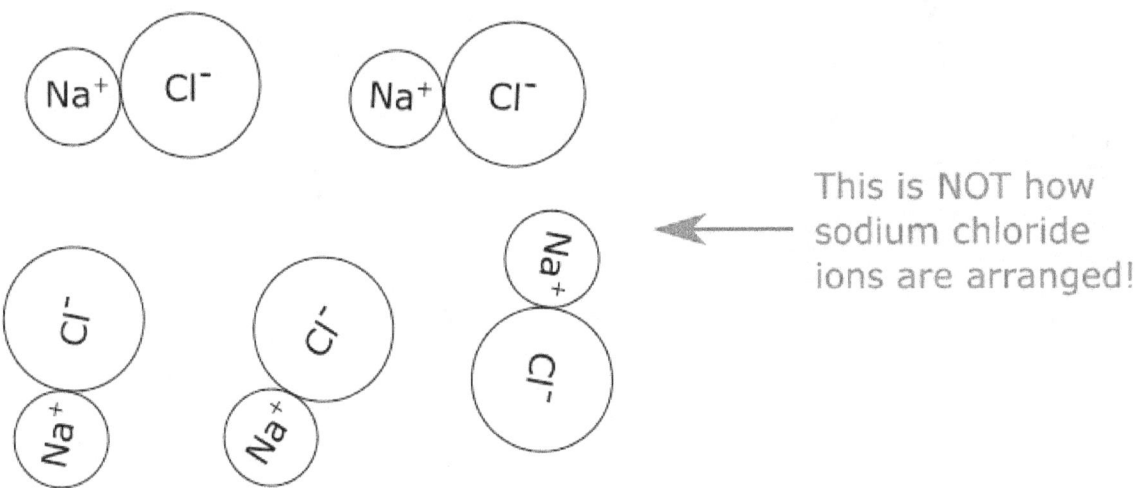

This is NOT how sodium chloride ions are arranged!

This is not how the ions are actually arranged.

The sodium and chloride ions are actually arranged (in solid form) as shown below:

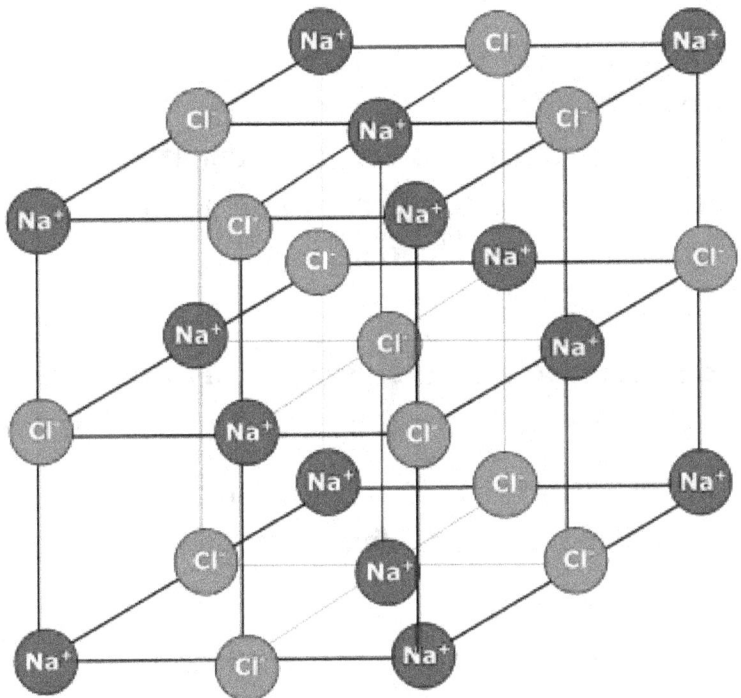

Each ion is surrounded by oppositely charged ions. The closest neighbours of a sodium ion are 6 chloride ions and the closest neighbours of a chloride ion are 6 sodium ions.

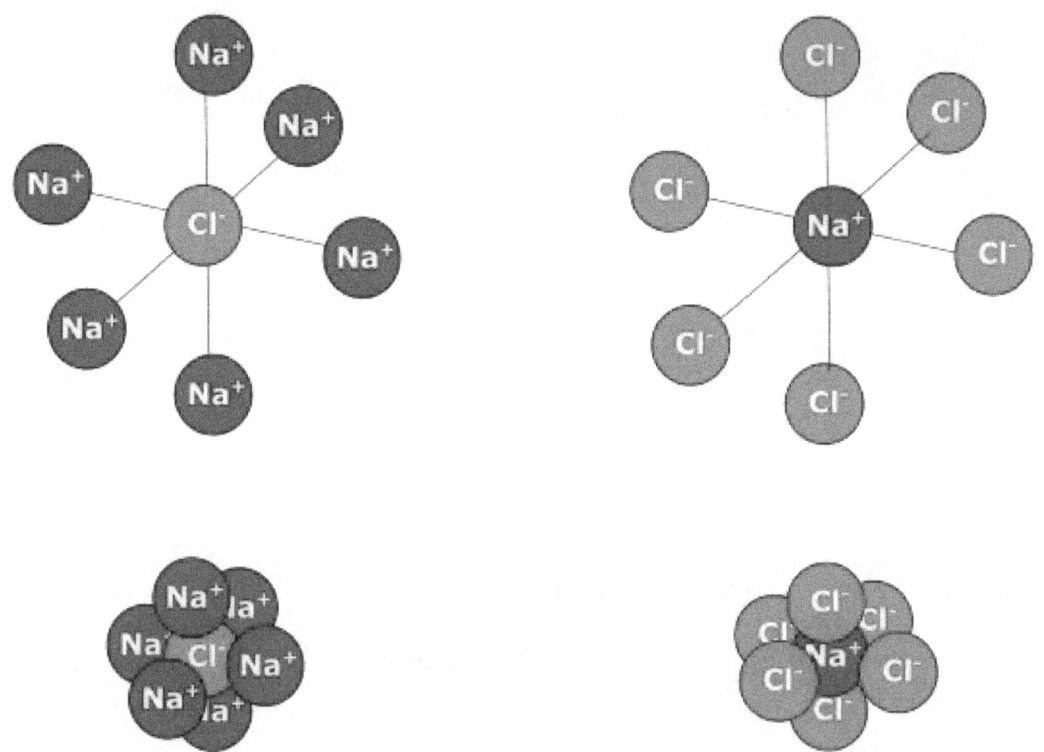

The picture above shows lines joining the sodium ions to chloride ions. These lines represent very strong electrostatic forces of attraction between the oppositely charged ions. In order to break these forces of attraction (for example, to melt a piece of sodium chloride solid), one has to supply an enormous amount of energy. That's why sodium chloride has a high melting point (801°C).

In general, **the ions in an ionic compound arrange themselves in a three-dimensional lattice structure** which makes it difficult to melt.

In solid form, the ions in the lattice structure are held in place and cannot move freely. Thus, **ionic compounds do not conduct electricity when they are solid.**

However, once an ionic compound is dissolved in water (and many ionic compounds are soluble in water), the lattice structure is destroyed and the ions are free to move. Hence, **dissolved ionic compounds are able to conduct electricity via their mobile ions.**

This brings us to one question: are ionic compounds soluble in water? **Most ionic compounds dissolve in water**, but not all.

Ionic compounds also conduct electricity when molten. Once an ionic compound melts, the lattice structure is again destroyed and the ions become mobile. Hence it can conduct electricity.

Test Your Understanding 2

For questions 1 to 5, circle the correct answer.

1. When a calcium atom is converted to a calcium ion

    A one electron is gained.
    B two protons are gained.
    C two electrons are lost.
    D two electrons are gained.

2. Which of the following pairs of elements is most likely to form an ionic compound?

    A Oxygen and chlorine
    B Sodium and oxygen
    C carbon and hydrogen
    D Helium and sodium

3. Which of the following properties best indicates that a substance is an ionic compound?
    A It is a solid at room temperature and pressure.
    B It is insoluble in water.
    C It conducts electricity when molten, but not when it is solid.
    D It can be synthesized from its elements.

4. Element X has the electronic structure 2, 8, 1 and element Y has the electronic structure 2, 7.
The compound formed between X and Y will probably

A be a liquid at room temperature and pressure.
B conduct electricity when it is in molten form.
C have a low boiling point.
D not dissolve in water.

5. Element X has a proton number of 3
Element Y has a proton number of 8
Element Z has a proton number of 10

Which one of the following is a correct statement regarding X, Y and Z?

A X and Z can combine together to form an ionic compound.
B Z is a Group VII element.
C X and Y can combine together to form an ionic compound.
D X gains an electron to form a negative ion.

6. The element lithium has a proton number of 3 and the element oxygen has a proton number of 8.

(a) In which Group of the Periodic Table is lithium placed?

(b) What is the charge of a lithium ion?

(c) Draw a 'dot and cross' diagram of the compound formed when lithium reacts with oxygen to form lithium oxide.

7. Three elements X, Y and Z have atomic (or proton) numbers between 8 and 13. Atom X has one electron more than a noble gas structure and atom Y has 3 electrons more than a noble gas structure. Atom Z, however, has 2 electrons fewer than a noble gas structure.

(a) Which element is most likely to have an ion carrying a
 (i) +3 charge?

 (ii) -2 charge?

(b) Complete the electronic structures of X, Y and Z.

 (i)   X: 2, 8, ____
 (ii) Y: 2, ____, ____
 (iii) Z: 2, ____

(c) Which of the elements X, Y, Z are metals?

(d) Use the Periodic Table at the end of this book to identify the elements and write down their chemical symbols in the blanks below.

 X: _____   Z: _____

 Y: _____

(e) Draw a 'dot and cross' diagram of the compound formed between Y and Z in the space below.
Use the actual chemical symbols for Y and Z from your answer in (d).

The Covalent Bond

We have seen that elements in the Periodic Table try to obtain a stable valence shell by either losing or gaining electrons. Metals form ionic bonds with non-metals by transferring one or more of their valence electrons to non-metals. These form oppositely charged ions which attract each other with strong electrostatic forces of attraction.

However, atoms also have another way to get a stable valence shell. **Non-metals can also form bonds with other non-metals using another method called covalent bonding.**

Consider a hydrogen and a chlorine atom as shown below.

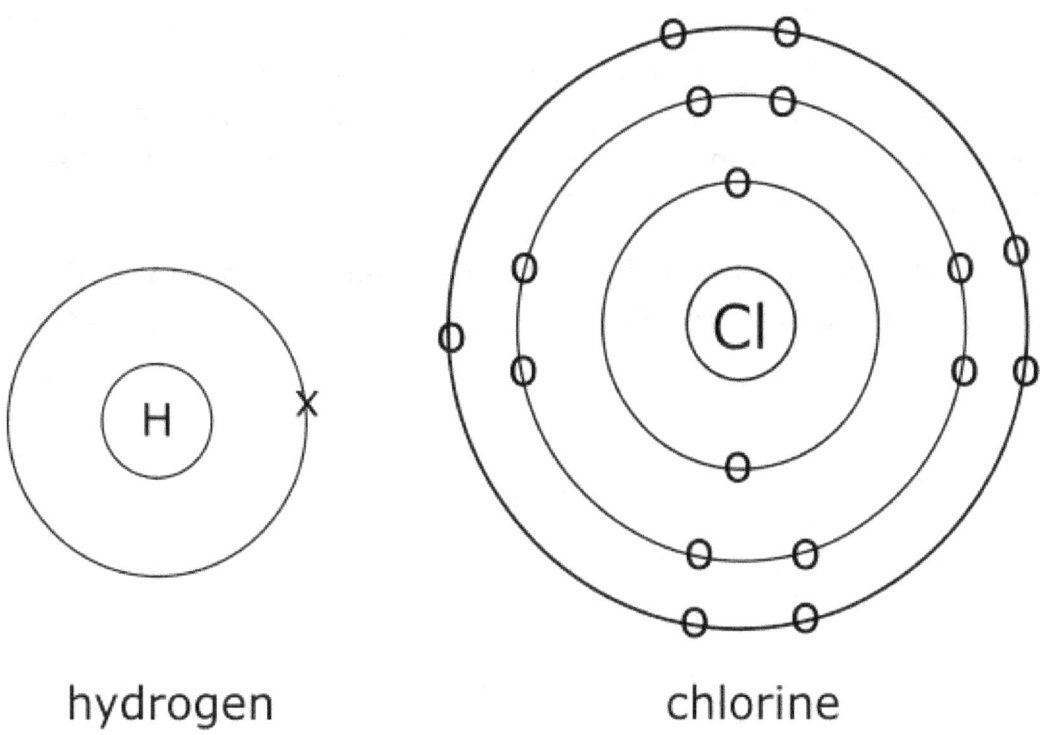

hydrogen   chlorine

The hydrogen atom needs two electrons in its valence (and only) shell to gain a stable configuration like helium while the chlorine atom needs just one more electron to have a stable valence shell of 8 electrons (like argon).

Both atoms do this by sharing a valence electron each:

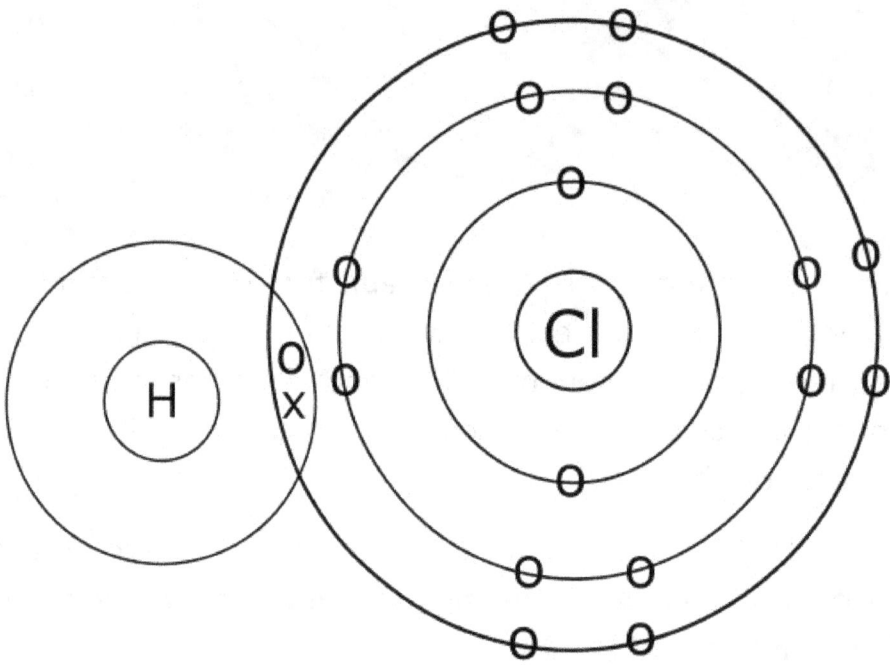

Look at what has happened. The hydrogen atom now has 2 valence electrons while the chlorine atom has the much desired 8 electrons in its valence shell! A new substance, hydrogen chloride, has been formed. This substance, the compound hydrogen chloride, is called a covalent compound and **the bond formed by the sharing of electrons is called the covalent bond**.

Note that both hydrogen and chlorine are non-metals and **covalent bonding is the mechanism that non-metals use to form bonds between themselves.**

The electrons in the other shells of the chlorine atom, i.e., the first and second shell don't contribute to the covalent bond. So, they are sometimes ignored when drawing the atoms:

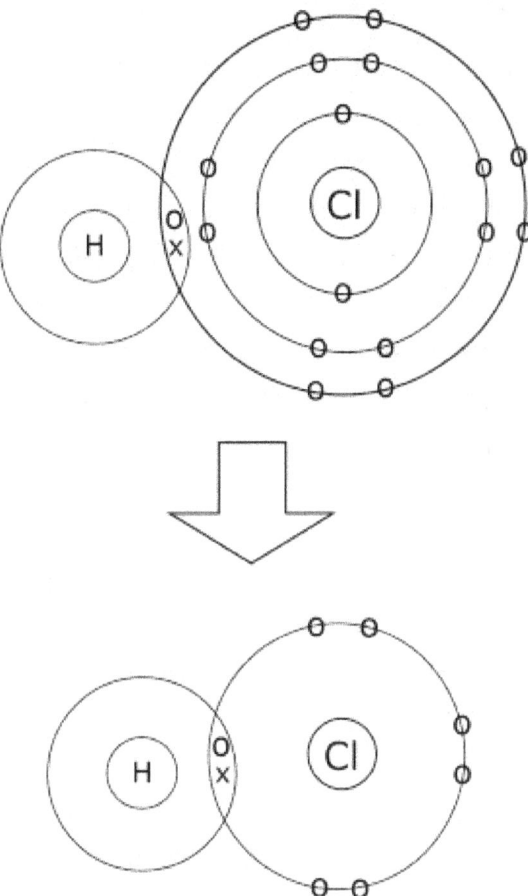

We call the combination of such covalently bonded atoms **molecules. A molecule refers to two or more covalently bonded atoms.**

Covalent bonds can also be formed between atoms of the same element. Chlorine atoms can also combine with each other like this:

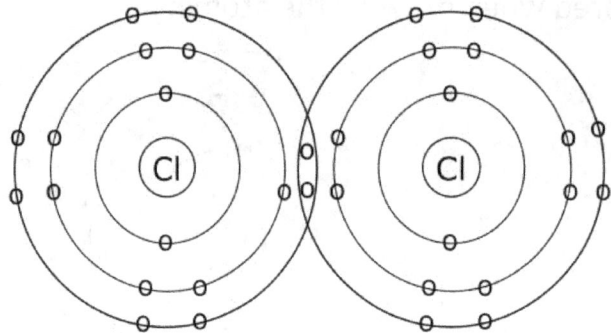

Each chlorine atom shares a valence electron with the other, forming a covalent bond between them. Ignoring the inner shells, the molecule of chlorine can be drawn like this:

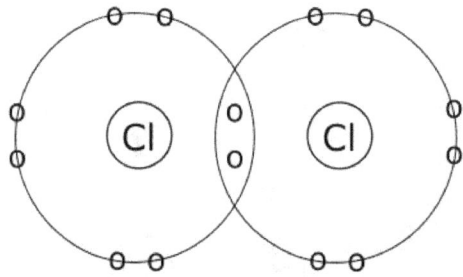

Hydrogen and fluorine atoms also pair up to form covalent molecules.

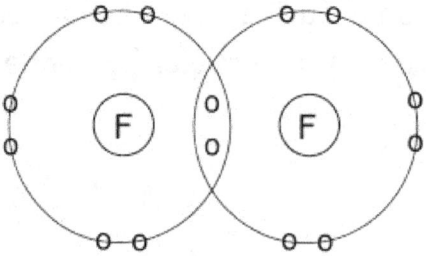
fluorine molecule

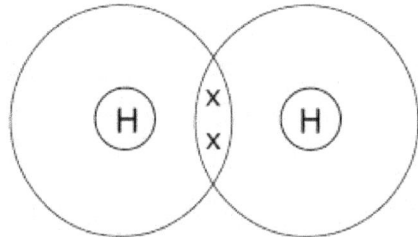
hydrogen molecule

Molecules formed from atoms of the same type are called molecules of elements. We have just discussed the molecules of hydrogen, fluorine and chlorine.

Molecules formed from atoms of different elements are called molecules of compounds. The hydrogen chloride molecule is an example of a molecule of a compound.

Single Covalent Bond Molecules

The hydrogen chloride molecule is one where each atom shares a single electron with another atom. This type of covalent bond is called a single bond. Fluorine and hydrogen molecules, seen in the previous page are also molecules formed from single bonds.

Another example of a single bond covalent molecule is the water molecule:

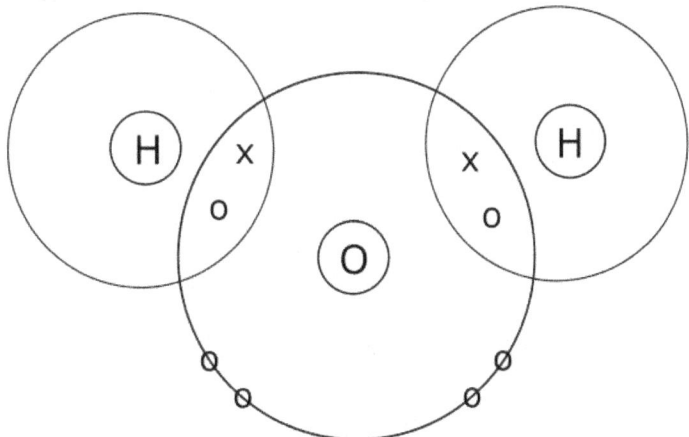

Only the valence shells are shown in the above diagram.

Double Covalent Bond Molecules

Atoms can also share more than one electron per atom.

Take the oxygen molecule. Each oxygen atom shares 2 of its valence electrons with each other to gain a stable valence shell:

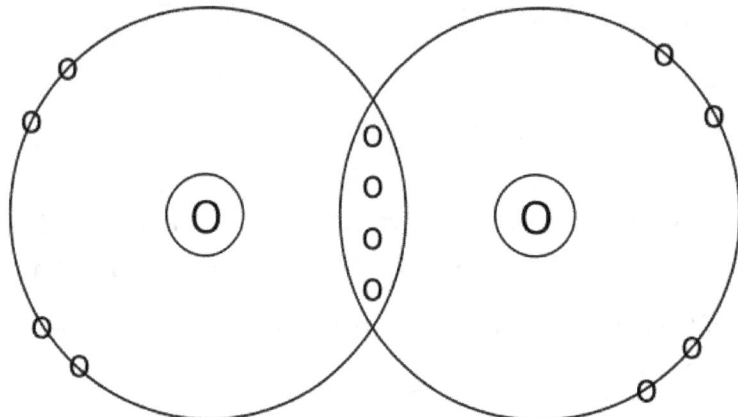

Some compounds contain a mixture of single and double bonds, for example, the ethene molecule:

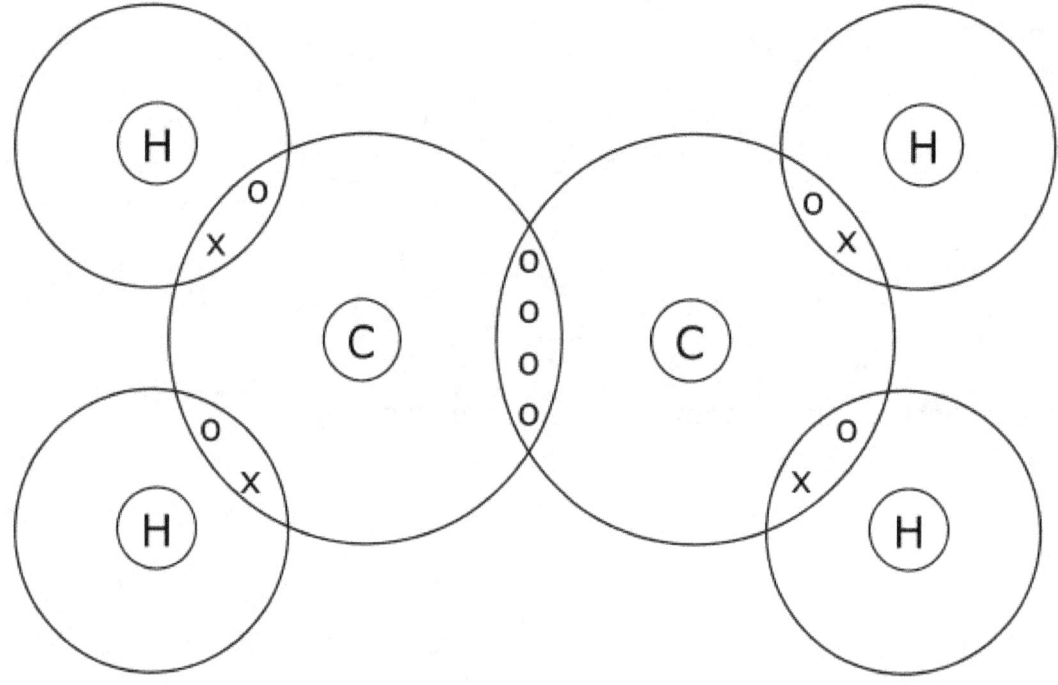

As seen above, the two carbon atoms form a double bond with each other and single bonds with each hydrogen atom.

Triple Covalent Bond Molecules

A triple covalent bond can form when atoms share 3 electrons with each other.

Consider the nitrogen molecule (only the valence shell electrons are shown):

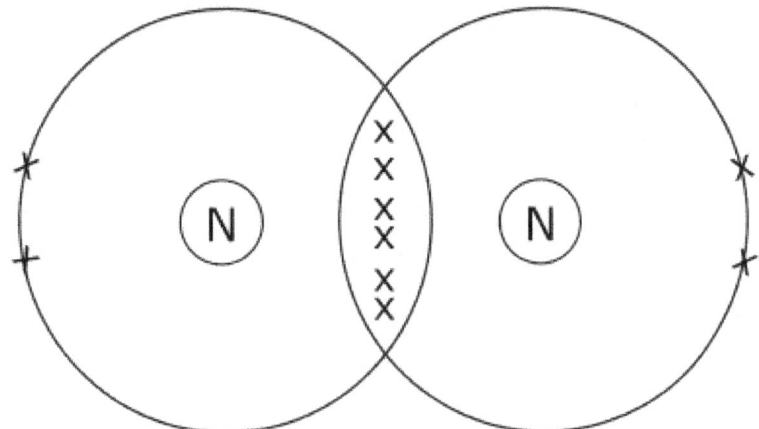

By sharing electrons this way, each nitrogen atom now has a stable valence shell of 8 electrons. Neat, isn't it?

Carbon is another atom that can form triple bonds. The diagram below shows a molecule of ethyne, a compound made of carbon and hydrogen.

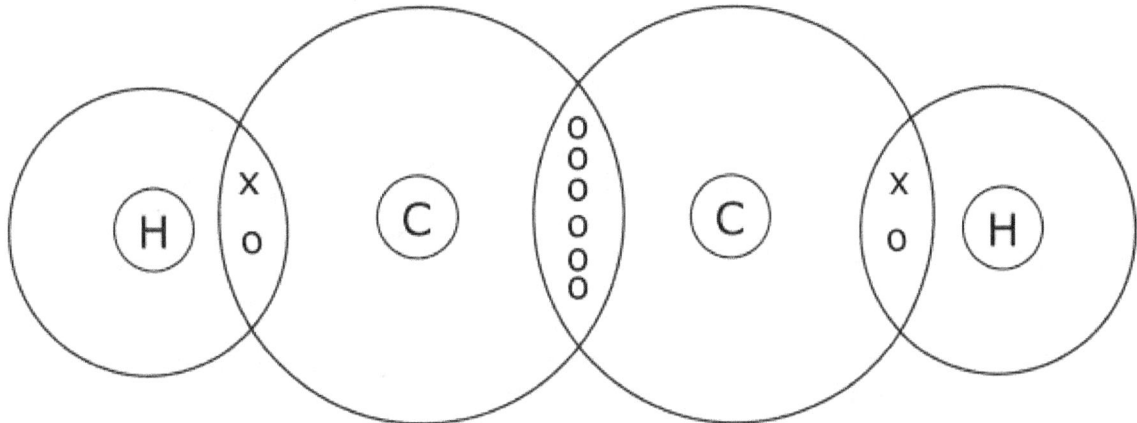

Simple Molecules

Many covalent substances exist as simple molecules.

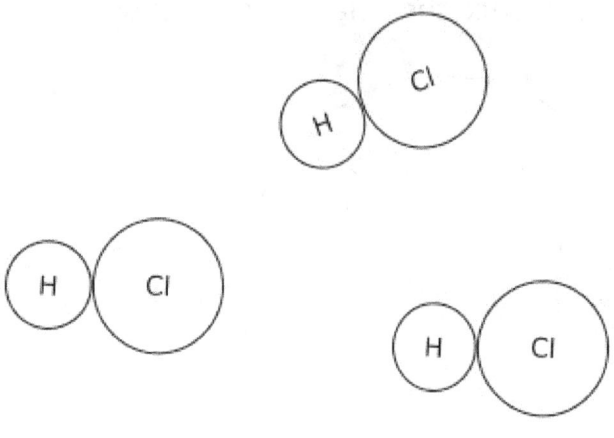

simple molecules

Each molecule is only attracted to other similar molecules by weak inter-molecular forces of attraction. The diagram below shows hydrogen chloride molecules.

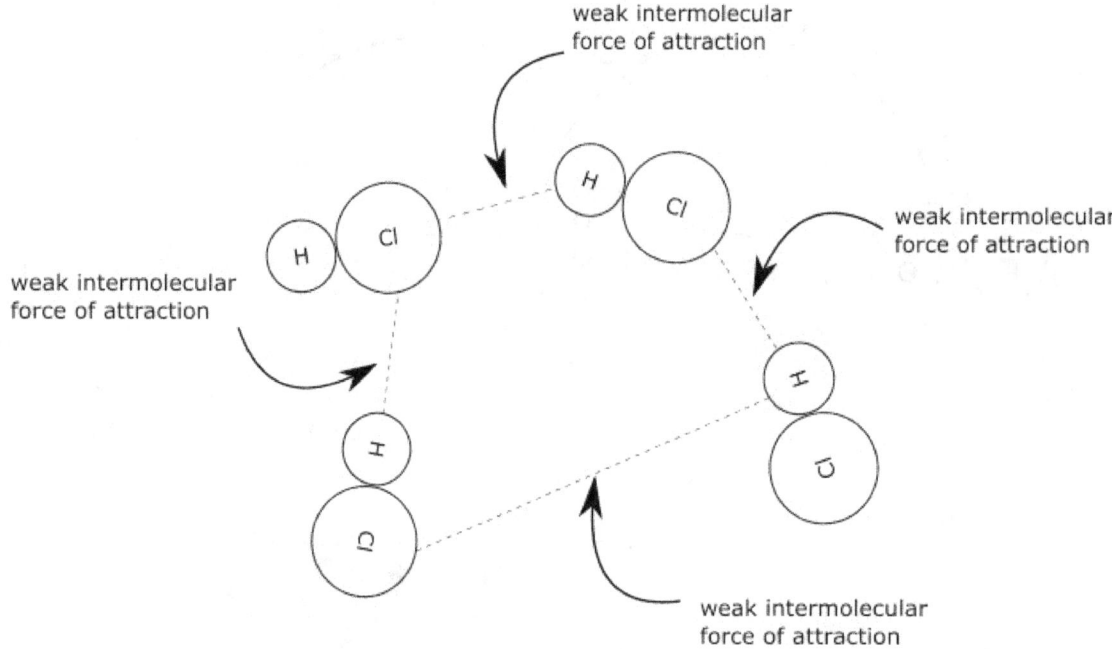

Because these weak inter-molecular forces of attraction can easily be broken without much energy, **covalent compounds which exist as simple molecules have low melting and low boiling points.**

Giant Covalent Structures

Many covalent compounds are liquids and gases at room temperatures. This is because they exist as simple molecules as discussed in the previous chapter.

One may be tempted to think that covalent bonds are weak but this is definitely not the case. **Covalent bonds are very strong**!

Consider the water molecule. One oxygen atom is bonded to two hydrogen atoms. You can heat this water molecule to a very, very high temperature and it will still remain a water molecule. You will not be able to separate the hydrogen atoms from the oxygen atoms easily. This is because the covalent bonds between hydrogen and oxygen are very strong. The reason why water has such a low melting and boiling point is because water exists as simple molecules and it is the weak inter-molecular forces of attraction between the molecules that can be easily broken by applying heat. This is why many covalent compounds are soft and some are liquids and gases.

Not all covalent compounds are soft, however. Look at diamond. It is a form of carbon and each carbon atom is joined to four other carbon atoms by very strong covalent bonds.

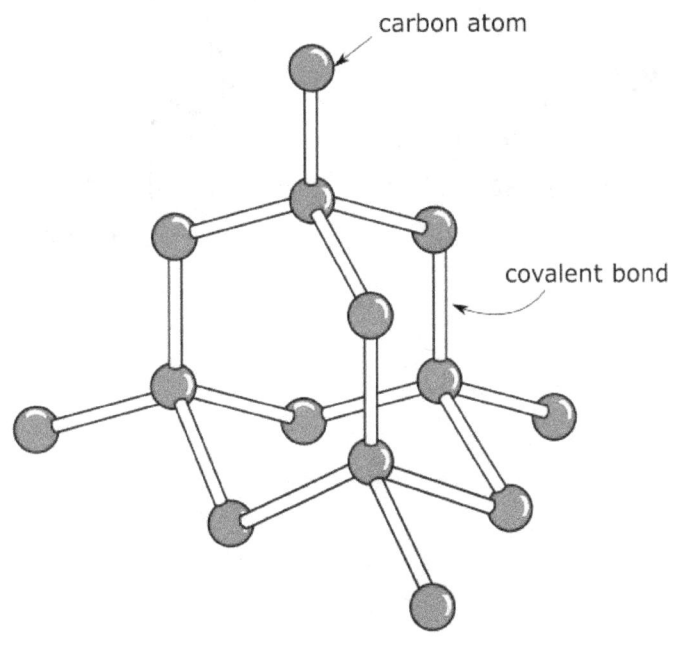

Diamond does not exist as simple molecules. Rather **the carbon atoms in diamond form a giant molecular structure**. Each carbon atom is joined to another by strong covalent bonds and not weak intermolecular forces of attraction. Thus, diamond is both very hard and has a high melting point.

Another covalent compound, silicon dioxide has a structure that is very similar to diamond. Below is a *simplified* diagram of the silicon dioxide structure.

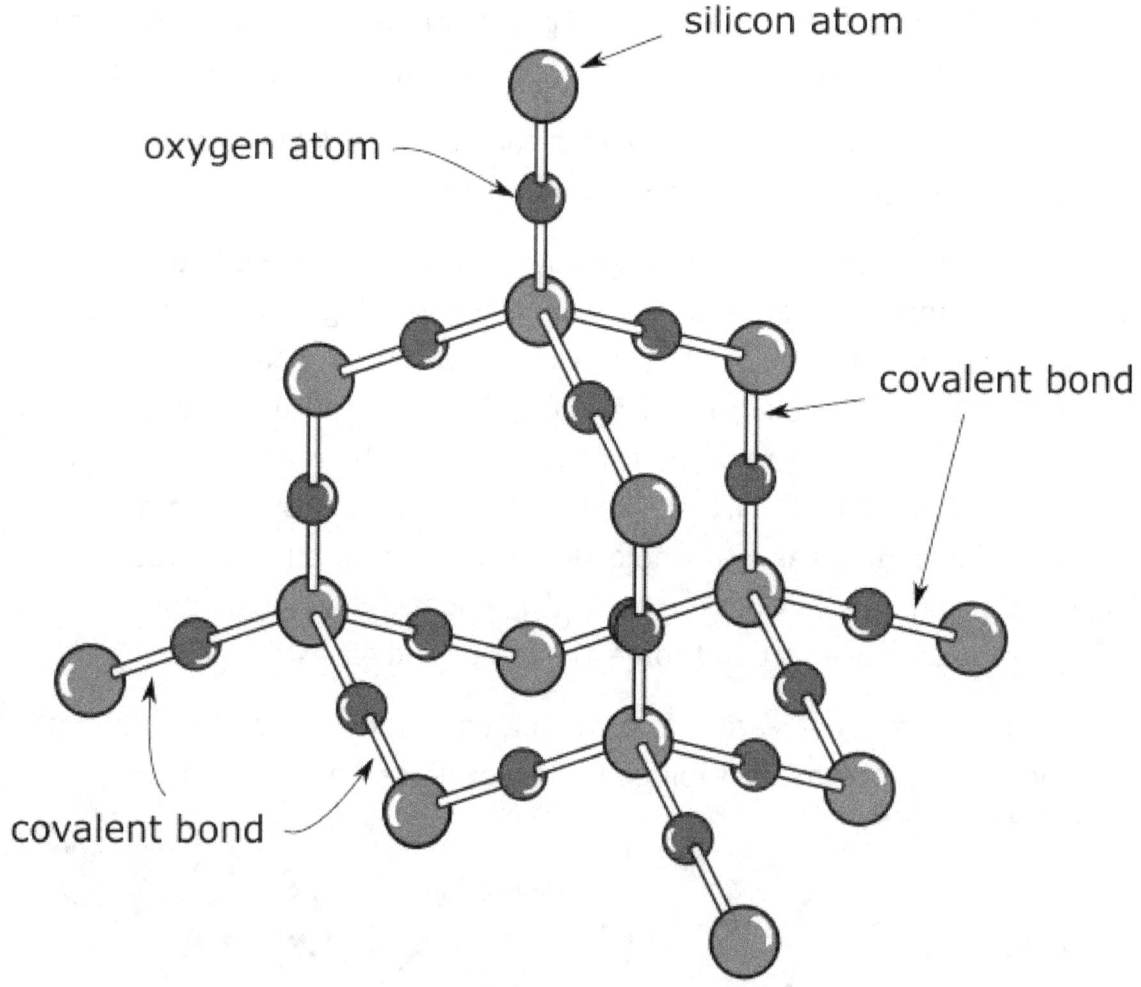

Silicon dioxide, also known as sand, has a giant molecular structure, just like diamond. And just like diamond, silicon dioxide has a very high melting point and is very hard.

Let us examine another giant molecular structure: graphite.

Now, graphite is another form of carbon and its structure is shown below:

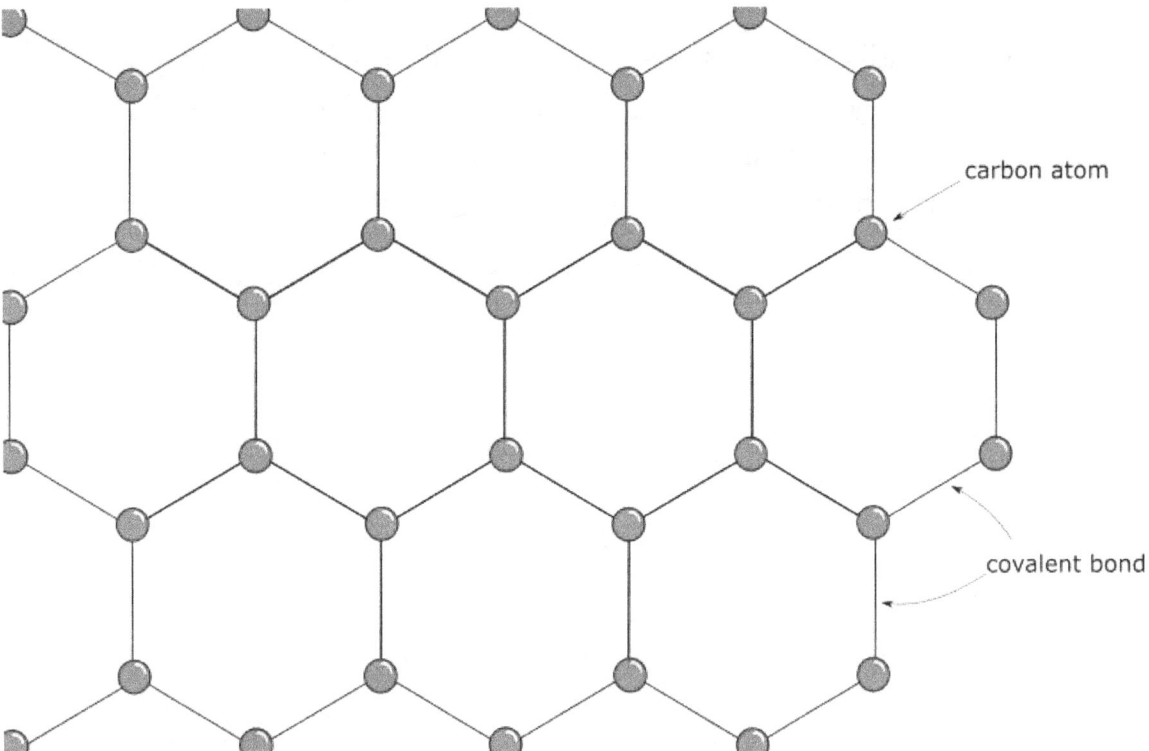

Each carbon atom in the graphite structure is covalently bonded to three other carbon atoms:

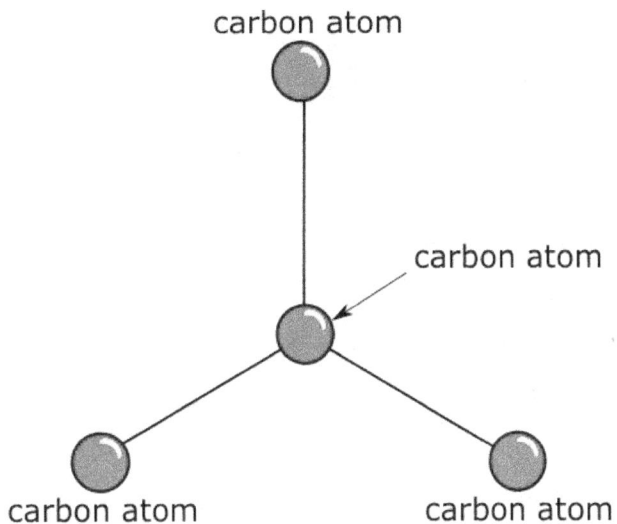

Carbon has 4 valence electrons but only three are used to form covalent bonds with three other carbon electrons. That means that there is one valence electron that is not used for bonding. This valence electron becomes detached from the carbon atom and becomes delocalized. It is free to move about and hence, graphite is able to conduct electricity in solid form.

Diamond, on the other hand, does not have any free electrons as all the 4 valence electrons of its carbon atoms are used for covalent bonding. Hence, diamond does not conduct electricity.

Compared to diamond's 3-dimensional structure, graphite's structure is 2-dimensional. The hexagonal arrangement of the graphite atoms is laid out on a single flat plane. A piece of graphite has many of these flat planes or layers stacked one on top of each other.

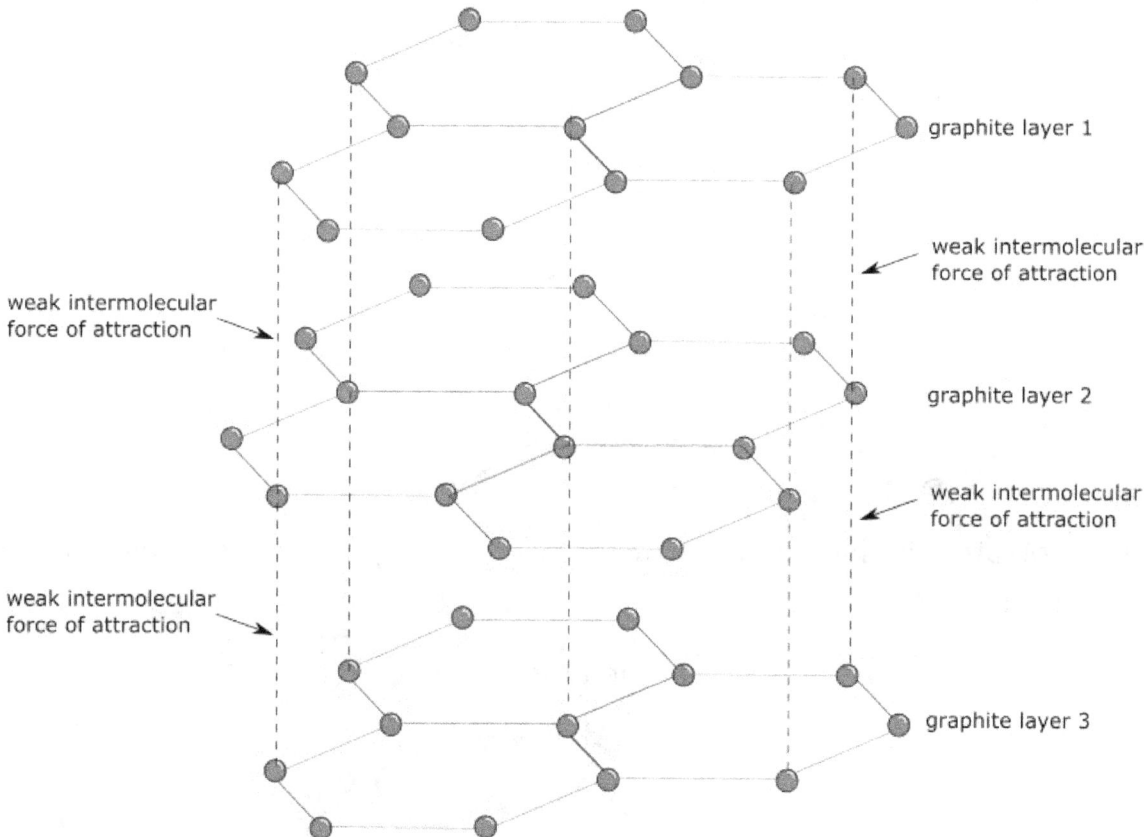

Weak intermolecular forces of attraction bind the layers together. But since these forces are easily overcome, the graphite layers slide over each other easily. Hence, solid graphite is nowhere near as hard as diamond and is a slippery substance, used as lubricants and pencil leads.

Did you notice something? When we were discussing ionic bonding, we left out Group IV elements. It is difficult for these atoms to form ions, but they readily form covalent bonds with themselves and other non-metals.

Comparing Ionic and Covalent Substances

We have touched upon some of the properties of ionic and covalent substances previously.

Let's summarize the differences in physical properties between these two types of substances.

Properties	Ionic Substances	Covalent Substances (Simple Molecules)	Covalent Substances (Giant Molecular Structures)
Solubility in Water	Many ionic substances are soluble in water (with some exceptions) but insoluble in organic solvents like kerosene.	Many covalent substances are insoluble in water but soluble in organic solvents (like kerosene)	Insoluble in all types of solvents.
Melting & Boiling Points	High melting and boiling points.	Low melting and boiling points	High melting and boiling points.
Electrical Conductivity	Cannot conduct electricity in solid form but can conduct electricity in aqueous (dissolved in water) and molten form.	Cannot conduct electricity in solid or molten form. Some covalent substances break up into ions when aqueous and can conduct electricity.	Cannot conduct electricity (except for graphite).

Writing Formulas of Ionic and Covalent Compounds

Ionic Compounds

To understand how to write the chemical formula of an ionic compound, we need to look at the ions that make up the compound. You have seen a few examples of that in the chapter on the Ionic Bond.

Consider sodium chloride. It consists of two oppositely charged ions, the sodium ion, Na^+ and the chloride ion Cl^- (chlorine ions are called chloride).

The Na^+ ion has a charge of +1 and the Cl^- ion has a charge of -1. So, the ratio should be 1:1 since the charges are equal but opposite and cancel each other out. So, we would write sodium chloride as Na_1Cl_1. But we don't need to write '1' in the formula, so it becomes NaCl.

What about a compound such as magnesium chloride? The magnesium ion has a +2 charge and the chloride, a -1 charge. So, we need two Cl^- ions to balance the +2 charge in the Mg^{2+} ion, i.e., $MgCl_2$.

Another way of looking at this is as follows:

1. Write down the charges of each ion side by side. Ignore the '+' or '-' signs.

    i.e., $Mg^2 \quad Cl^1$

2. Swap the numbers and write them as subscripts below each atom:

    i.e., $Mg_1 \quad Cl_2$

3. There you have it: Mg_1Cl_2 which becomes $MgCl_2$.

Let's consider a slightly more difficult example: aluminium oxide.

Looking at the Periodic Table, we find that aluminium has a +3 charge while oxygen as a -2 charge. Oxygen ions are called oxide, by the way.

1. Write the ions, sans the signs, beside each other:

    $Al^3 \quad O^2$

2. Swap the numbers: $\quad Al_2 \quad O_3$

3. Your formula would be Al_2O_3

What about a compound such as iron (II) oxide?

The (II) in the formula refers to the charge on the iron ions, i.e., Fe^{2+}. As we know from the previous two examples, the oxide ion is O^{2-}. If we follow the steps above, we get the chemical formula Fe_2O_2. Since both Fe and O have the same subscript, we can omit them and write the formula as FeO.

Notice that the chemical symbol for the metal is written first, followed by the non-metal. So, sodium chloride is NaCl and not ClNa and Iron (II) oxide is FeO, not OFe.

There is a group of ions, made up of more than one atom, which behave like single ions. They are usually non-metallic ions chemically combined together for form a positively or negatively charged ion.

For example, the nitrate ion NO_3^-, is made of nitrogen and oxygen combined together in the ratio shown and has a net -1 charge. Such ions are called polyatomic ions (or radicals in older chemistry literature). Examples of such polyatomic ions are:

NH_4^+ (ammonium ion), SO_4^{2-} (sulfate ion), CO_3^{2-} (carbonate ion) and OH^- (hydroxide ion)

How do we write the formula of a compound involving these polyatomic ions?

Consider a compound such as calcium hydroxide.

1. Write the calcium ion (and its charge sans the sign) and the hydroxide ion next to it.

    Ca^2 OH^1

2. Swap the numbers.

    Ca_1 $(OH)_2$

Did you notice the brackets surrounding the hydroxide ion? This is important. As the hydroxide ion is behaving like a single ion, the '2' subscript means that there is a pair of OH ions, not just the hydrogen as would be the case if the brackets were not used.

3. The formula of calcium hydroxide is thus $Ca(OH)_2$.

Covalent Compounds

The components in a covalent compound do not exist as ions, but when writing the formulas for covalent compounds, we can use the same method as that for writing the formulas of ionic compounds ... up to a point. This may help you get an *idea* of the chemical formula of the compound but because covalent bonds between two or more non-metallic elements can have many different variations – single, double or triple bond, etc – the formula obtained may not be the correct one.

Consider two non-metallic elements, say carbon and hydrogen. Carbon has 4 valence electrons, so we can give it a valency of 4, while hydrogen has 1 valence electron, so it has a valency of 1. We can write the elements as

$$C^4 \quad H^1$$

Using the method that we used for ionic compounds, we swap the valency numbers:

$$C_1H_4$$

Thus, we have the compound CH_4. This is methane, which is a gas at room temperature. But covalent compounds made of carbon and hydrogen can also have formulas such as C_2H_2, C_2H_4, C_2H_6, C_8H_{18}, etc. All these are very different compounds from methane!

Furthermore, for covalent compounds, you may not simplify the subscripts to the simplest form. For example, the covalent compound ethyne, C_2H_2 cannot have its subscripts simplified to CH. No such compound exists.

It is not just with carbon and hydrogen. Many non-metallic elements combine in different ratios and form completely different compounds.

So, for covalent substances, one has to be careful when trying to figure out the formulas of the compounds.

Because of the nature of sharing electrons between atoms, we can represent covalent compounds in a number of different ways.

Look at ammonia. Its chemical formula is NH$_3$. Its structure looks like this:

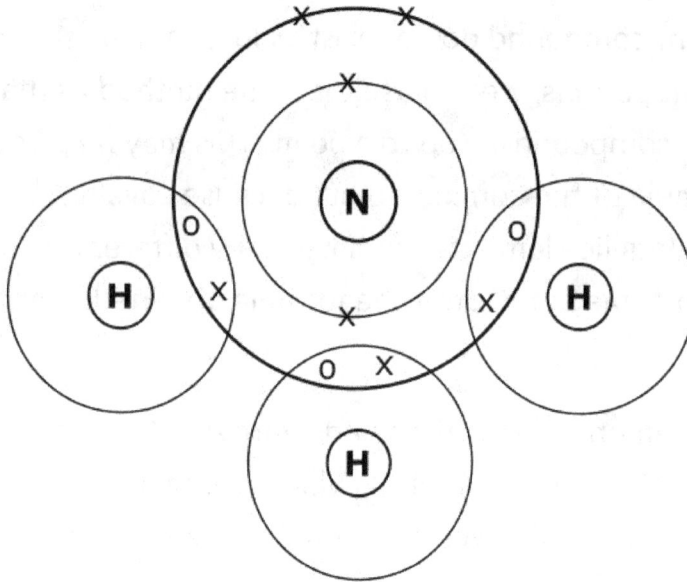

The above 'dot and cross' diagram can be more conveniently drawn as

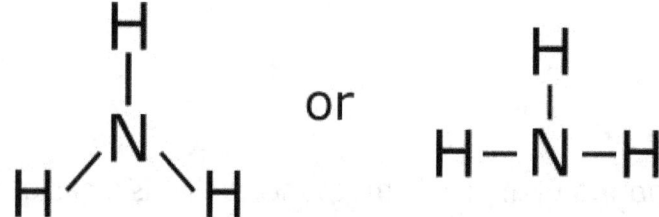

This representation is called the **structural formula** of ammonia. Compare the structural formula with the 'dot and cross' diagram:

1. Each line represents a single covalent bond, which is essentially a pair of shared electrons.
2. The structural formula does not show the innermost electrons. In fact, it doesn't show electrons at all.

Let's examine another covalent molecule.

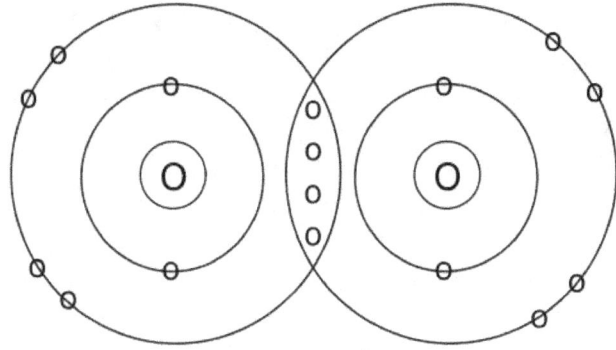

The molecular formula for the above molecule is O_2.

The structural formula is $O=O$

In the previous section, it was stated that a single line represents a pair of shared electrons. Since there is a double bond holding the oxygen atoms together, i.e., 2 pairs of electrons (i.e., 4 electrons), 2 lines (=) are used.

Example Question

A molecule of carbon dioxide CO_2, has the structural formula. $O=C=O$.

(a) How many of electrons in the molecule are used for bonding?
(b) How many electrons in the molecule are **not** used for bonding?

Answer

(a) There is a double bond between the carbon atom and each oxygen atom. Each double bond involves 4 electrons. There are 2 double bonds in the molecule. Hence the total number of electrons used for bonding is 2 x 4 = **8**.

(b) Electronic configuration of carbon is 2, 4. All the 4 valence electrons are used in bonding but the inner shell 2 electrons are not.
Electronic structure of oxygen is 2, 6. Four valence electrons and two inner shell electrons are not used. So, each oxygen atom has 6 electrons not used for bonding, giving a total of 12 unused electrons for both oxygen atoms.
Hence total number of electrons not used for bonding = 12 + 2 = **14**.

Test Your Understanding 3

For questions 1 to 5, circle the correct answer.

1. Which one of the following compounds is covalent?

    A sulfur dioxide
    B sodium sulfide
    C calcium chloride
    D potassium iodide

2. Element X has a proton number of 5.
   Element Y has a proton number of 6.
   Element Z has a proton number of 8.

   Which of the following is a correct statement regarding X, Y and/or Z?

    A X and Y can combine together to form a covalent compound.
    B X and Z can combine together to form a covalent compound.
    C Y and Z can combine together to form an ionic compound.
    D Y and Z can combine together to form a covalent compound.

3. In the table below, the electronic configurations of four elements are shown. The letters representing the elements are not their actual chemical symbols.

Element	Electronic Configuration
W	2, 8,8
X	2,6
Y	2,4
Z	2,8,2

   Which of the following shows the correct formula for a covalent compound formed from the elements in the table?

   A WX   B YX$_2$ C Z$_2$X   D ZY

4. The diagram below shows the bonding in nitric acid.

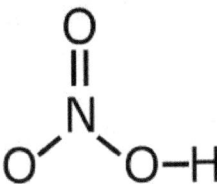

What is the total number of electrons in the covalent bonds surrounding the nitrogen atom?

A 4   B 6   C 8   D 16

5. The diagram below shows a molecule containing the elements X, Y and Z.

Which elements (chemical symbols) could X, Y and Z be?

	X	Y	Z
A	C	H	Br
B	H	C	Br
C	Br	C	H
D	C	H	K

6. Element X forms a compound with chlorine with the formula XCl_3. The compound has a melting point of -93.6°C and a boiling point of 76.1°C.

State, with reasons,

(a) the physical state (solid, liquid or gas) of XCl_3 at room conditions (about 24°C and 1 atmosphere),

(b) the type of bonding in XCl$_3$.

(c) whether X is likely to be a metal or non-metal.

7. Hydrogen sulfide is a smelly gas at room conditions.

(a) State whether hydrogen sulfide is an ionic or covalent compound. Give a reason for your answer.

(b) The chemical formula of hydrogen sulfide is H$_2$S. Draw a 'dot and cross' diagram of hydrogen sulfide, showing ALL the electrons.

8. Difluoromethane is a compound that is used in industries.

   Its molecular formula is CH_2F_2.

   Its structural formula is

   ```
       H
       |
   F − C − F
       |
       H
   ```

   (a) Draw a 'dot and cross' diagram of a difluoromethane molecule in the space below.
   Show only the valence electrons.

   (b) Difluoromethane has a melting point of − 136°C and a boiling point of - 51.7°C.
   Explain why this compound has such low melting and boiling points.

Metallic Bonding

Atoms of metals tend to lose their valence electrons easily.

Thus, a bunch of metallic atoms will look something like this:

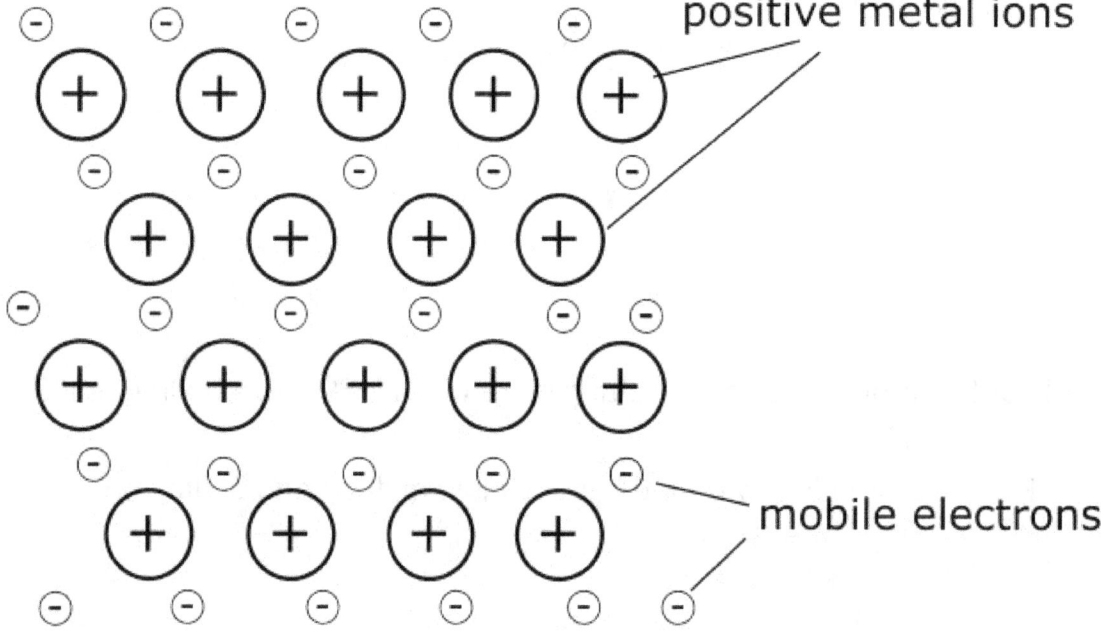

The strong force of attraction between the positive ions and the negative electrons constitutes the metallic bonds.

In solid metals, the positive ions are held in a giant lattice structure surrounded by a sea of negatively charged mobile electrons.

These electrons no longer belong to any positive ion and are called 'delocalized' electrons. They are free to move (i.e., mobile) around the metallic lattice and hence can conduct electricity.

Properties of metals

1. Metallic bonds are very strong bonds. A lot of energy is required to break them. Hence metals usually have very high melting and boiling points.

2. Because of the strong attraction between oppositely charged ions, the particles are tightly packed together, making metals very dense.

3. As the electrons in metals are mobile, metals can conduct electricity in both solid and molten states.

Hybrid Bonding

Many compounds are made of both ionic and covalent components. Consider the compound calcium carbonate, $CaCO_3$. It is made up of the calcium ion, Ca^{2+} and the carbonate ion, CO_3^{2-}. The bond between the calcium ion and the carbonate ion is ionic, but the carbonate ion is covalent in nature. Its structure looks like this:

$$\left[O=C \begin{smallmatrix} O \\ O \end{smallmatrix} \right]^{2-}$$

(arrows pointing to: double covalent bond; single covalent bond)

The carbonate ion is an example of a **polyatomic** ion, i.e., it is made up of more than two atoms.

Another example of hybrid bonding is sodium nitrate, $NaNO_3$. It is made up of an ionic bond between the sodium Na^+ ion and the nitrate NO_3^- ion, but within the nitrate ion, the bonds are covalent:

$$\left[O=N \begin{smallmatrix} O \\ O \end{smallmatrix} \right]^{-}$$

(arrows pointing to: double covalent bond; single covalent bond)

Can you think of any other compounds made up of such hybrid (ionic and covalent) bonds?

Practice Questions

For questions 1 to 5, circle the correct answer.

1 Which of the following has the same number of protons as the carbonate ion, CO_3^{2-}? Use the Periodic Table to help you find the answer.

   A Zn   B Kr   C Ca^{2+} D SO_4^{2-}

2 Which of the following pairs of particles have equal and opposite charges?

   A A hydrogen ion and a proton
   B A hydrogen ion and a neutron
   C A proton and an electron
   D A proton and a neutron

3 The element with electronic configuration 2, 8, 7

   A forms an ionic compound with potassium.
   B is a metal.
   C forms an ion of charge +1.
   D reacts only with non-metals.

4 In which of the following pairs do the elements form a compound by sharing electrons?

   A neon and argon   B fluorine and sodium
   C carbon and chlorine   D calcium and oxygen

5   The diagram below shows two atoms X and Y.

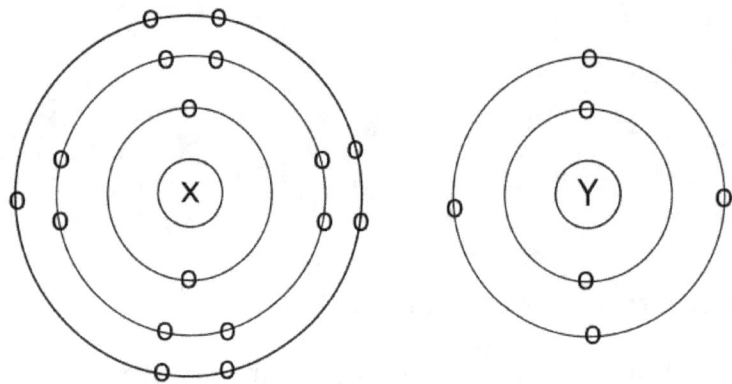

Which of the following correctly describes a compound made of X and Y?

	Chemical formula	Ionic or covalent?
A	XY	Ionic
B	XY	Covalent
C	YX$_4$	Ionic
D	YX$_4$	Covalent

6. A chemical compound, formed by the reaction between chlorine and magnesium has a melting point of 714°C and a boiling point of 1412°C.

(a) Is the compound solid, liquid or gas at room temperature (24°C)?

(b) Write down the chemical formula of the compound.

(c) Is the compound ionic or covalent? Explain your answer.

(d) In the space below, draw a 'dot and cross' diagram of the compound formed from magnesium and chlorine.

7   Carbon dioxide, CO_2 and silicon dioxide SO_2 are both covalent compounds. At -78.5°C, carbon dioxide changes state from solid to gas without going through a liquid state. Silicon dioxide however, melts at 1710°C and boils at 2230°C.

(a) Explain why these two compounds exhibit such a big difference in the temperatures at which they change state.

(b) Draw a single molecule of carbon dioxide, showing clearly the covalent bonding between the atoms. You need only draw the valence shells.

Answers to Questions

Test Your Understanding 1

1 (a)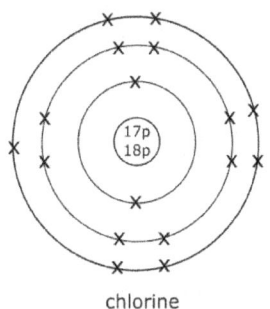

chlorine

(b) Cl⁻

(c) An ion of chlorine has 18 electrons.

(d) argon

2 (a)

Sub-atomic particles	Number
Proton	12
Neutron	24
Electron	12

(b) Magnesium has 2 valence electrons.

(c) Mg^{2+}

(d) To get the stable electronic configuration of neon.

55

3

Element	Electronic configuration / structure of atom	Symbol of ion formed	Name of ion	Electronic configuration / structure of ion	Noble gas whose configuration of ion is similar to
Calcium	2,8,8,2	Ca^{2+}	calcium ion	2,8,8	Argon
Fluorine	2,7	F^-	fluoride ion	2,8	Neon
Lithium	2,1	Li^+	llthium ion	2	Helium
Chlorine	2,8,7	Cl^-	chloride ion	2,8,8	Argon
Nitrogen	2,5	N^{3-}	Nitride ion	2,8	Neon
Potassium	2,8,8,1	K^+	Potassium ion	2,8,8	Argon

Test Your Understanding 2

1	2	3	4	5
C	B	C	B	C

6 (a) Group I (b) +1

(c)

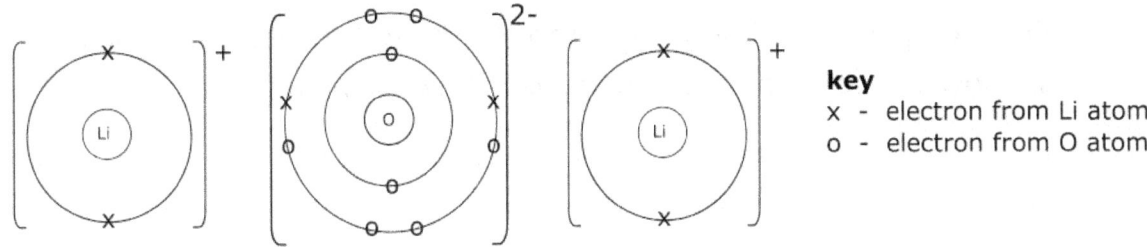

7 (a) (i) Y (ii) Z

(b) (i) X: 2, 8, 1 (ii) Y: 2, 8, 3 (iii) Z: 2, 6

(c) X and Y are metals (metals have 3 or fewer valence electrons)

(d) X: Na   Z: O

   Y: Al

(e)

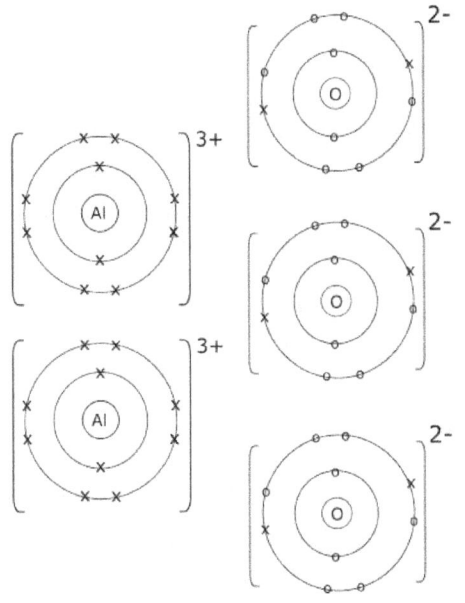

Test Your Understanding 3

1	2	3	4	5
A	D	B	C	A

6 (a) liquid   (b) Covalent bonding   (c) Non-metal

7 (a) covalent compound; As it is a gas at room temperature, its boiling point is quite low and covalent compounds have low boiling points.

(b)

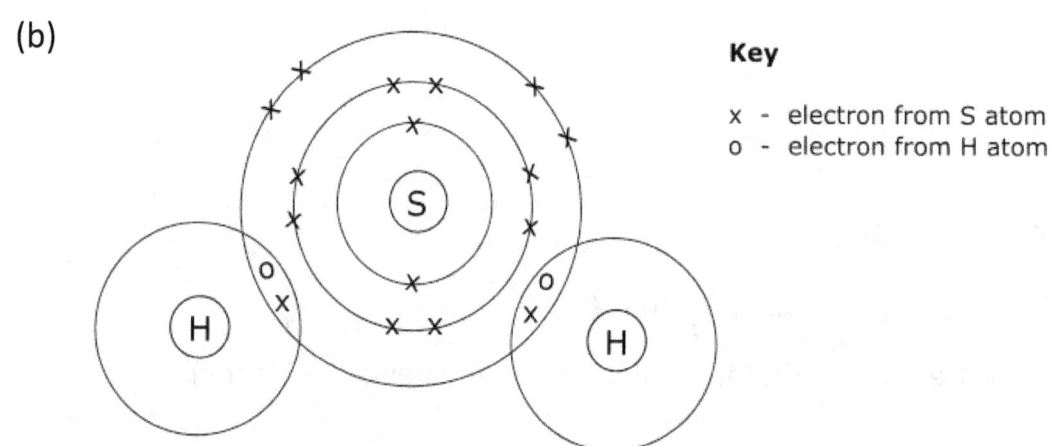

Key

x - electron from S atom
o - electron from H atom

8 (a)

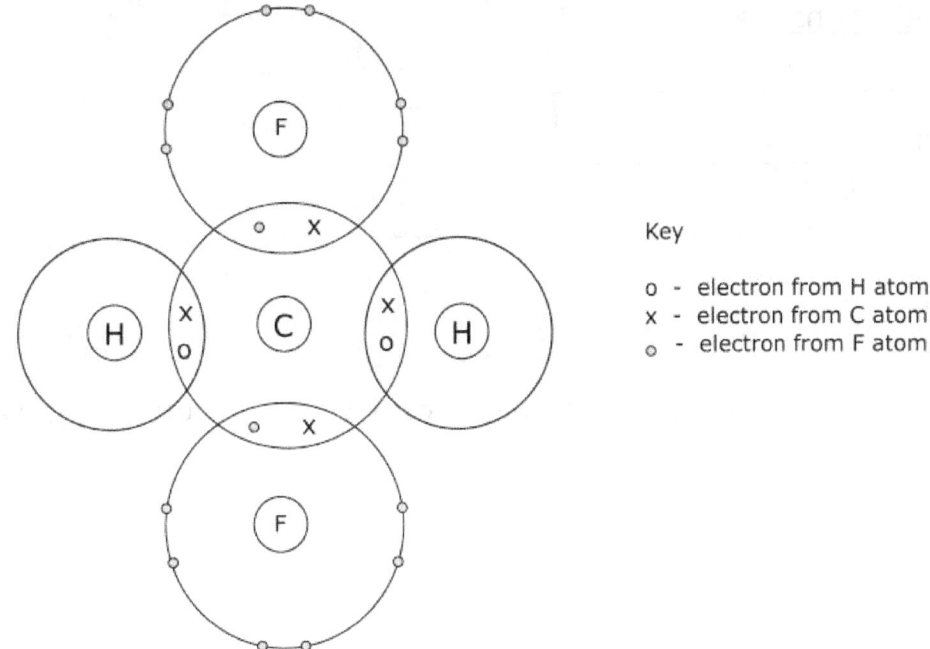

Key

o - electron from H atom
x - electron from C atom
o - electron from F atom

(b) Difluoromethane exists as simple covalent molecules. The weak intermolecular forces of attraction between the molecules requires little energy to break.

Practice Questions

1	2	3	4	5
A	C	A	C	D

6 (a) Solid   (b) $MgCl_2$

(c) It is an ionic compound because it is formed from a metal ion and a non-metal ion.

(d)
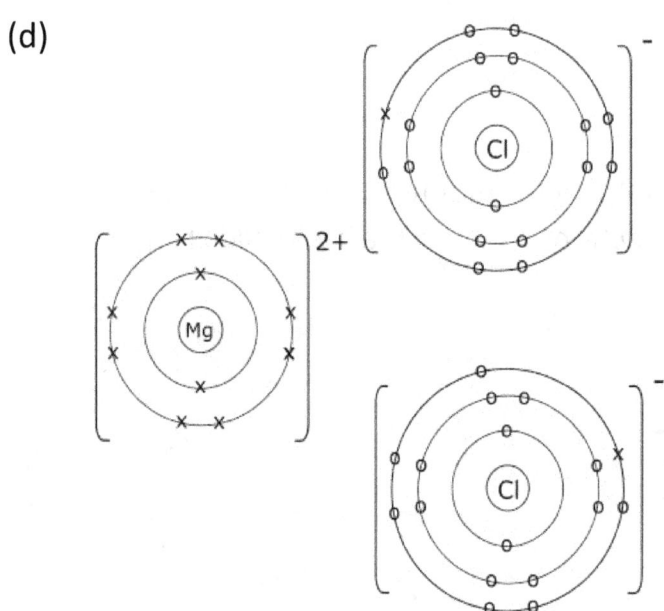

Key
o - electron from Cl atom
x - electron from Mg atom

7 (a) Carbon dioxide exists as simple molecules and the weak inter-molecular forces of attraction between the molecules require little energy to overcome. Silicon dioxide however, exists as a giant molecular structure and the strong covalent bonds between the atoms require a lot of energy to overcome.

(b)

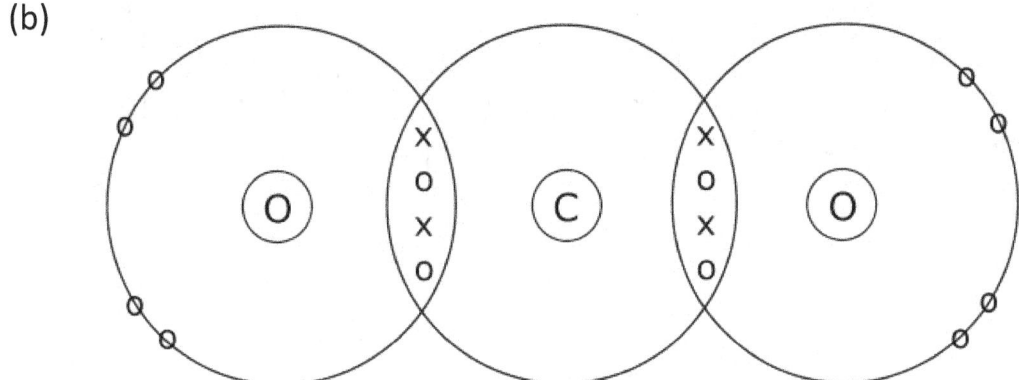

THE PERIODIC TABLE OF ELEMENTS

I	II												III	IV	V	VI	VII	0
						1 H Hydrogen 1												2 He Helium 4
3 Li Lithium 7	4 Be Beryllium 9												5 B Boron 11	6 C Carbon 12	7 N Nitrogen 14	8 O Oxygen 16	9 F Fluorine 19	10 Ne Neon 20
11 Na Sodium 23	12 Mg Magnesium 24												13 Al Aluminium 27	14 Si Silicon 28	15 P Phosphorus 31	16 S Sulfur 32	17 Cl Chlorine 35.5	18 Ar Argon 40
19 K Potassium 39	20 Ca Calcium 40	21 Sc Scandium 45	22 Ti Titanium 48	23 V Vanadium 51	24 Cr Chromium 52	25 Mn Manganese 55	26 Fe Iron 56	27 Co Cobalt 59	28 Ni Nickel 59	29 Cu Copper 64	30 Zn Zinc 65		31 Ga Gallium 70	32 Ge Germanium 73	33 As Arsenic 75	34 Se Selenium 79	35 Br Bromine 80	36 Kr Krypton 84
37 Rb Rubidium 85	38 Sr Strontium 88	39 Y Yttrium 89	40 Zr Zirconium 91	41 Nb Niobium 93	42 Mo Molybdenum 96	43 Tc Technetium -	44 Ru Ruthenium 101	45 Rh Rhodium 103	46 Pd Palladium 106	47 Ag Silver 108	48 Cd Cadmium 112		49 In Indium 115	50 Sn Tin 119	51 Sb Antimony 122	52 Te Tellurium 128	53 I Iodine 127	54 Xe Xenon 131
55 Cs Caesium 133	56 Ba Barium 137	57 - 71 lanthanoids	72 Hf Hafnium 178	73 Ta Tantalum 181	74 W tungsten 184	75 Re Rhenium 186	76 Os Osmium 190	77 Ir Iridium 192	78 Pt Platinum 195	79 Au Gold 197	80 Hg Mercury 201		81 Tl Thallium 204	82 Pb Lead 207	83 Bi Bismuth 209	84 Po Polonium -	85 At Astatine -	86 Rn Radon -
87 Fr Francium -	88 Ra Radium -	89 - 103 actinoids	104 Rf Rutherfordium -	105 Db Dubnium -	106 Sg Seaborgium -	107 Bh Bohrium -	108 Hs Hassium -	109 Mt Meitnerium -	110 Ds Darmstadtium -	111 Rg Roentgenium -	112 Cn Copernicium -			114 Fl Flerovium -		116 Lv Livermorium -		

Lanthanoids

57 La Lanthanum 139	58 Ce Cerium 140	59 Pr Praseodymium 141	60 Nd Neodymium 144	61 Pm Promethium -	62 Sm Samarium 150	63 Eu Europium 152	64 Gd Gadolinium 157	65 Tb Terbium 159	66 Dy Dysprosium 163	67 Ho Holmium 165	68 Er Erbium 167	69 Tm Thulium 169	70 Yb Ytterbium 173	71 Lu Lutetium 175

Actinoids

89 Ac Actinium -	90 Th Thorium 232	91 Pa Protactinium 231	92 U Uranium 238	93 Np Neptunium -	94 Pu Plutonium -	95 Am Americium -	96 Cm Curium -	97 Bk Berkelium -	98 Cf Californium -	99 Es Einsteinium -	100 Fm Fermium -	101 Md Mendelevium -	102 No Nobelium -	103 Lr Lawrencium -

www.ingramcontent.com/pod-product-compliance
Lightning Source LLC
Chambersburg PA
CBHW080526220526
45465CB00006B/2610